Cambridge IGCSE™

Chemistry

Fourth Edition

Bryan Earl
Doug Wilford

Orders: please contact Hachette UK Distribution, Hely Hutchinson Centre, Milton Road, Didcot, Oxfordshire, OX11 7HH. Telephone: +44 (0)1235 827827. Email education@hachette.co.uk Lines are open from 9 a.m. to 5 p.m., Monday to Friday. You can also order through our website: www.hoddereducation.co.uk

ISBN: 978 1 3983 1050 6

© Bryan Earl and Doug Wilford 2021

First edition published in 2002,
Second edition published in 2009,
Third edition published in 2014,
This edition published in 2021 by
Hodder Education,
An Hachette UK Company
Carmelite House
50 Victoria Embankment
London EC4Y 0DZ

www.hoddereducation.com

Impression number   10 9 8 7 6 5 4 3 2 1

Year   2025 2024 2023 2022 2021

Cover photo © Björn Wylezich - stock.adobe.com

Illustrations by Integra Software Services Pvt. Ltd., Pondicherry, India.

Typeset in ITC Officina Sans Std Book 11.5/13 by Integra Software Services Pvt. Ltd, Pondicherry, India

Printed in Slovenia

A catalogue record for this title is available from the British Library.

Contents

Acknowledgements

The authors would like to thank Irene, Katharine, Michael and Barbara for their patience, support and encouragement throughout the production of this textbook. We would also like to thank the editorial and publishing teams at Hodder Education who have supported us on the journey over the past year.

Source acknowledgements
pp. 13, 38, 40, 41, 42, 189, 195, 196, 198, 199, 200 and 213

The molecular models shown were made using the Molymod® system available from Molymod® Molecular Models, Spiring Enterprises Limited, Billingshurst, West Sussex RH14 9NF England.

Photo credits
r right, *l* left, *t* top, *m* middle, *b* bottom
p.1 © *bl* Dolphinartin/stock.adobe.com, *tr* © 12ee12/stock.adobe.com, *br* © Olaf Schubert/Image Broker/Alamy Stock Photo; **p.2** *l* © Donald L Fackler Jr/Alamy Stock Photo, *r* © Andrew Lambert Photography/Science Photo Library; **p.3** *tl* © Geoscience Features Picture Library/Dr.Booth, *bl* © Javier Larrea/Agefotostock/Alamy Stock Photo, *br* © Power And Syred/Science Photo Library; **p.7** *l* © Andrew Lambert Photography/Science Photo Library, *r* © Photiconix/Alamy Stock Vector; **p.8** *l* © Andrew Lambert Photography/Science Photo Library, *r* © Andrew Lambert Photography/Science Photo Library; **p.10** © NASA/ESA/STSCI/Hubble Heritage Team/ Science Photo Library; **p.11** *t* © Robert Harding/ Getty Images, *m* © FlemishDreams – Fotolia, *b* © Echo 23 Media/Alamy Stock Photo; **p.12** *tl* © Bay Ismoyo/AFP/Getty Images, *bl* © Christie's Images / The Bridgeman Art Library, *tr* © BL Images Ltd / Brian Lawrence / Alamy Stock Photo; **p.13** © Andrew Lambert Photography/Science Photo Library; **p.15** © Andrew Lambert Photography / Science Photo Library; **p.18** *tr* © Galyna Andrushko – Fotolia, *br* © Andrew Lambert Photography / Science Photo Library; **p.19** © Andrew Lambert Photography/ Science Photo Library; **p.20** *(left column)* *tl* © Vvvita/stock.adobe.com, *tr* © Weyo/stock.adobe. com, *bl* © Shutterstock/Stay_Positive, *bm* © Magraphics/stock.adobe.com, *br* sommai/stock. adobe.com, *(right column)* *tr* © Science Photo Library/Eye of Science, *br* © moodboard/Thinkstock; **p.21** *tr* © Digital Instruments/Veeco/Science Photo Library; **p.24** *br* © Henry Westheim Photography / Alamy Stock Photo; **p.30** *tl* © Andrew Lambert Photography/ Science Photo Library, *tr* © Andrew Lambert Photography/Science Photo Library, *b* © midosemsem – Fotolia; **p.33** © Science Source/ Science Photo Library; **p.36** *l* © Martyn F. Chillmaid, *r* © Martyn F. Chillmaid; **p.37** *(left column)* *l* © Martyn F. Chillmaid, *r* © Martyn F. Chillmaid, *(right column)* Courtesy of the University of Illinois at Urbana-Champaign Archives; **p.38** © Andrew Lambert Photography/Science Photo Library; **p.40** © Andrew Lambert Photography/Science Photo Library; **p.41** *l* © Andrew Lambert Photography/Science Photo Library, *r* © Andrew Lambert Photography/ Science Photo Library; **p.42** © Andrew Lambert Photography/Science Photo Library; **p.44** © E.R.Degginger/Science Photo Library; **p.46** *l* © Philippe Plailly/Science Photo Library, *r* © Robert Harding Photo Library; **p.48** *tl* © Richard Megna/ Fundamental/Science Photo Library, *bl* © Sheila Terry/Science Photo Library, *tr* © Michael Pettigrew – Fotolia, *br* © Kage Mikrofotografie GBR/Science Photo Library; **p.49** © overthehill – Fotolia; **p.56** *l* © Andrew Lambert Photography/Science Photo Library, *r* © Andrew Lambert Photography/Science Photo Library; **p.57** © Andrew Lambert Photography/ Science Photo Library; **p.63** © Dirk Wiersma/Science Photo Library; **p.69** *tl* © Alexander Maksimenko – Fotolia, *tr* © Mkos83/stock.adobe.com, *b* © Pavel Losevsky/stock.adobe.com; **p.72** © Howard Davies/ Alamy Stock Photo; **p.73** © Kekyalyaynen/ Shutterstock.com; **p.74** *l* © dpa picture alliance archive / Alamy Stock Photo, *r* © Rex Features; **p.75** *l* © Kathy Gould/iStock/Thinkstock, *r* © Trevor Clifford Photography/Science Photo Library; **p.78** *tr* © Andrew Lambert Photography/Science Photo Library, *br* © Clynt Garnham Renewable Energy / Alamy Stock Photo; **p.81** © Frankie Martin/NASA/ Science Photo Library; **p.82** © Adam Hart-Davis/ Science Photo Library; **p.83** *bl* © Andrew Lambert Photography/Science Photo Library, *br* © Andrew Lambert Photography/Science Photo Library; **p.87** *tl* © Dinodia Photos / Alamy Stock Photo, *bl* © Oleg Zhukov – Fotolia, *mt* © Matt K – Fotolia, *mb* © Dmitry Sitin – Fotolia, *tr* © iofoto – Fotolia, *br* © yang yu – Fotolia; **p.88** *l* © Andrew Lambert Photography/Science Photo Library, *r* © Sindo Farina/Science Photo Library; **p.89** © Stevie Grand/ Science Photo Library; **p.90** © GeoScience Features Picture Library; **p.101** *tl* © Illustrez-vous - Fotolia, *bl* © Arnaud Santinil – Fotolia, *tr* © Yakovlevadaria/ stock.adobe.com, *br* © National Motor Museum, Motoring Picture Library / Alamy Stock Photo;

How to use this book

To make your study of Chemistry for Cambridge IGCSE™ as rewarding and successful as possible, this textbook, endorsed by Cambridge Assessment International Education, offers the following important features:

FOCUS POINTS

Each topic starts with a bullet point summary of what you will encounter within the topic.

This is followed by a short outline of the topic so that you know what to expect over the next few pages.

Test yourself

These questions appear regularly throughout the chapter so you can check your understanding as you progress.

Revision checklist

At the end of each chapter, a revision checklist will allow you to recap what you have learned in each topic and double-check that you understand the key concepts before moving on.

Exam-style questions

Each chapter is followed by exam-style questions to help familiarise learners with the style of questions they may see in their examinations. These will also prove useful in consolidating your learning. Past paper questions are also provided at the end of the book.

As you read through the book, you will notice that some text is shaded yellow. This indicates that the highlighted material is Supplement content only. Text that is not shaded covers the Core syllabus. If you are studying the Extended syllabus, you should look at both the Core and Supplement sections.

Practical skills

These boxes identify the key practical skills you need to understand as part of completing the course.

Key definitions

These provide explanations of the meanings of key words as required by the syllabus.

Worked examples

These boxes give step-by-step guidance on how to approach different sorts of calculations, with follow-up questions so you can practise these skills.

Going further

These boxes take your learning further than is required by the Cambridge syllabus so that you have the opportunity to stretch yourself.

Answers are provided online with the accompanying *Cambridge IGCSE Chemistry Teacher's Guide*. A *Practical Skills Workbook* is also available to further support you in developing your practical skills as part of carrying out experiments.

Scientific enquiry

Throughout your IGCSE Chemistry course, you will need to carry out experiments and investigations aimed at developing some of the **skills** and **abilities** that scientists use to solve real-life problems.

Simple experiments may be designed to measure, for example, the temperature of a solution or the rate of a chemical reaction. Longer experiments, or investigations, may be designed to allow you to actually see the relationship between two or more physical quantities such as how rate of reaction varies with temperature and concentration.

Investigations are likely to come about from the topic you are currently studying in class, and your teacher may provide you with suggestions. For all investigations, both your teacher and you must consider the safety aspects of the chemicals and apparatus involved. You should never simply carry out a chemistry investigation without consideration of the hazards of the chemicals or getting the approval of your teacher.

To carry out an investigation, you will need to:

1 **Select and safely use suitable techniques, apparatus and materials** – your aim must be to safely collect sufficient evidence using the most appropriate apparatus for the technique you have chosen. Being able to draw and label diagrams correctly to show how the equipment will be used is also important. Your techniques will need to be explained clearly to do a proper risk assessment. For example, how to carry out a titration, how you are going to follow the rate of the reaction you are using, or how to test for ions and gases.

2 **Plan your experiment** – this is an important part of doing science and involves working out what you are going to do to try to find answers to the questions you have set yourself. Predictions based on work you have been studying or are doing in class may help you develop the investigation in terms of the number and type of observations or data needed. You will also need to be able to identify the independent and dependent variables. For example, if you are trying to find out how temperature affects the rate of a reaction, the temperature will be the independent variable, but the dependent variable might be the volume of gas collected. Other variables such as concentration need to be controlled so that they will not affect the data

obtained. Most importantly, it is essential that you carry out a risk assessment before you do any practical work.

3 **Make and record observations** – the data you need to answer the questions you have set yourself can only be found if you have planned your investigation sensibly and carefully. For example, you might start to use a measuring cylinder to collect a gas, but as you develop your ideas you may realise a burette might be more appropriate and more accurate. Be careful not to dismantle the equipment/apparatus until you have completed your analysis of the data, and you are sure you do not need to repeat any of the measurements! If you have to reset your equipment/apparatus, it may add further errors to the results you have obtained.

Ensure that all your data, numeric or observational, is displayed in a clear format. This will often be in the form of headed tables with the correct units being shown to the appropriate degree of precision.

4 **Interpret and evaluate observations** – the results you obtain from any investigation must be displayed carefully and to the accuracy of the equipment you have chosen to use. Your choice of presentation will help you interpret your evidence and make conclusions. Often your presentation will be in the form of a graph or a table. For some graphs, you may need to calculate gradients or use it to find values at a specific point during the investigation by drawing intercepts. Good chemists keep looking at the data and alter the way in which it is obtained to get more accurate results. You should be able to evaluate whether your data is good or bad. If it is good, were there any anomalous results? Why did you get them?

5 **Evaluate methods and suggest possible improvements** – at the end of your investigation, you must be able to evaluate the equipment, methods and techniques that you have used. Think about any sources of errors that could have affected your results by the use of the wrong equipment. Consider, if you were able to carry out the investigation again, what you would change. The more data you obtain, the easier it is to spot anomalous results.

A **written report** of any chemical investigation would normally be made of these fixed components:

» First, state the **aim** of the work at the very beginning to inform your teacher what you were doing the investigation to find out.

» A list of all items of equipment/apparatus used and a record of the smallest division of the scale of each measuring device you have used (see Chapter 14). For example, burettes can be read to two decimal places, to the nearest $0.05\,cm^3$, where the second decimal place is either a 0 if the bottom of the meniscus is on the scale division, or a 5 if it is between the divisions. If the meniscus was between $24.10\,cm^3$ and $24.20\,cm^3$, the reading would be $24.15\,cm^3$.

» You must show that you have considered the safety of yourself and others before you carry out any practical work. Provide a list of all the chemicals you will use, as well as the ones you will produce, and do a **risk assessment** to check on all the hazards of the chemicals. The results of your risk assessment might indicate that you need to work in a well-ventilated room or in a fume cupboard. In some cases you may need the assistance of your teacher. If in doubt, always ask your teacher for advice.

» Clearly state the details of the **methods used**, starting with the wearing of eye protection. The methods should be shown as numbered steps and should be made as clear as possible. Ideas of the number of measurements that will be made and their frequency should be stated. Observations should be clear and you should use changes in colour and physical state as part of your observations.

» Presentation of **results** and **calculations**. If you made several measurements of a quantity, draw up a table in which to record your results. Use the column headings, or start of rows, to name exactly what the measurement is and state the units used; for example, in a rates of reaction experiment, 'Mass of calcium carbonate/g'.

Give numeric values to the number of significant figures appropriate to the equipment being used, for example, a mass could be recorded to 0.5 g or 0.05 g depending on the resolution of the top-pan balance you use. Take averages and remember that anomalous or non-concordant results should not be used in their calculation. If you decide to make a graph of your results, you will need at least six data points taken over as large a range as possible; be sure to label each axis of a graph with the name and unit of the quantity being plotted. Make sure that the scale you use allows the points to fill up as much of the graph paper as possible.

Clearly explain the calculations involved in the interpretation of your data and give the significant figures appropriate to the equipment used.

» **Conclusions** can be obtained from the graphs and calculations you carry out. Your conclusions from the data obtained might be different from those that you expected. Even so it is very important for any scientist to come to terms with the findings of their experimental result, good or bad!

» In the **evaluation** you should make a comparison between the conclusions of your investigation and your expectations: how close or how different were they? You should comment on the reliability and accuracy of the observations and the data obtained. Could you have improved the method to give better or more accurate results? Would a pH probe have been better than using universal indicator to find the point of neutralisation in a titration? Were there any anomalous points on your graphs, or any unusual data or observations? Highlight these and try to give an explanation.

Suggestions for investigations

Some suggested investigations are outlined in this book as follows:

1 Find which vinegars contain the most acid. (Chapter 8)
2 Find the molar volume of hydrogen by reacting magnesium with hydrochloric acid. (Chapter 4)
3 Determination of the enthalpy of combustion of ethanol. (Chapter 6)
4 The effect of changing the surface area of a reactant on the rate of reaction. (Chapter 7)
5 Use the anion and cation methods of identification to find the ions present in tap water. (Chapter 14)
6 Show that ammonia is a weak base by measuring its pH and conductivity, and comparing your results with those from a solution of sodium hydroxide with the same concentration. (Chapter 8)
7 Determine the melting point of stearic acid. (Chapter 1)
8 How can sodium chloride be obtained from rock salt? (Chapter 14)

9 What are the effects of acid rain on a variety of building materials? (Chapter 11)

10 Which is the best temperature, between 34 and 40°C, for the fermentation of sugar to take place? (Chapter 12)

11 What are the chemical properties of the weak organic acid, ethanoic acid? (Chapter 13)

12 Do foodstuffs contain carbon? (Chapter 12)

Ideas and evidence in science

In some of the investigations you perform in the school laboratory, you may find that you do not interpret your data in the same way as your friends do; perhaps you will argue with them as to the best way to explain your results and try to convince them that your interpretation is right. Scientific controversy frequently arises through people interpreting evidence differently.

For example, our ideas about atoms have changed over time. Scientists have developed new models of atoms over the centuries as they collected new experimental evidence. If we go back to the Greeks in the 5th century BC, they thought matter was composed of indivisible building blocks which they called atomos. However, the idea was essentially forgotten for more than 2000 years. Then John Dalton published his ideas about atoms in 1800. He suggested that all matter was made of tiny particles called atoms, which he imagined as tiny spheres that could not be divided. It then took another 100 years before Joseph Thomson, Ernest Rutherford and James Chadwick carried out experiments and discovered that there was a structure within the atom. This saw the continuous development of what we know today as atomic theory.

A further example involves the well-known Russian chemist Dimitri Mendeleev. He realised that the physical and chemical properties of the known elements were related to their atomic mass in a 'periodic' way, and arranged them so that groups of elements with similar properties fell into vertical columns in his table. However, in devising his table, Mendeleev did not conform completely to the order of atomic mass, with some elements swapped around. It took time for his ideas to gain acceptance because the increase in atomic mass was not regular when moving from one element to another. We now know, with the development of atomic theory and a better understanding of chemical processes, that the elements in the Periodic Table are not all in atomic mass order. It took until 1934, with an understanding of atomic number and post-Russian revolution, for the Periodic Table to be finally accepted in the form you see today.

There are many different types of scientists with specialties in their own areas of work such as chemistry and physics, but they all work in the same way. They come up with new theories and ideas, they carry out work to find the evidence to establish whether their ideas are correct and, if not, why. Scientists rely on other scientists checking their work, often improving the ideas of everyone and moving science forward. The use of new ideas is often beneficial to everyone in the world, for example, the discovery of vaccines for Covid-19, or the push to improve battery manufacture for use in electric cars which would in turn help solve one of the biggest problems we have to face: global warming. Scientists are working hard to stop global warming but their ideas are not always embraced because of economic and political factors.

1 States of matter

FOCUS POINTS
★ What is the structure of matter?
★ What are the three states of matter?
★ How does kinetic particle theory help us understand how matter behaves?

In this first chapter, you will look at the three states of matter: solids, liquids and gases. The structure of these states of matter and how the structures can be changed from one to another is key to understanding the states of matter.

You will use the kinetic particle theory to help explain how matter behaves, so you can understand the difference in the properties of the three states of matter and how the properties are linked to the strength of bonds between the particles they contain. Why, for example, can you compress gases but cannot compress a solid? By the end of this chapter, you should be able to answer this question and use the ideas involved to help you to understand many everyday observations, such as why car windows mist up on a cold morning or why dew forms on grass at night.

1.1 Solids, liquids and gases

Chemistry is about what **matter** is like and how it behaves, and our explanations and predictions of its behaviour. What is matter? This word is used to cover all the substances and materials from which the physical universe is composed. There are many millions of different substances known, and all of them can be categorised as solids, liquids or gases (Figure 1.1). These are what we call the three states of matter.

b Liquid

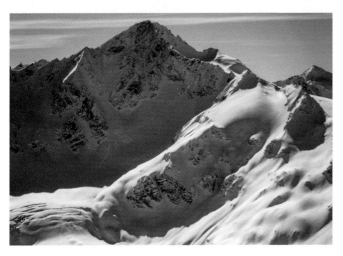

a Solid

c Gas

▲ **Figure 1.1** Water in three different states

A **solid**, at a given temperature, has a definite volume and shape which may be affected by changes in temperature. Solids usually increase slightly in size when heated, called **expansion** (Figure 1.2), and usually decrease in size if cooled, called **contraction**.

A **liquid**, at a given temperature, has a fixed volume and will take up the shape of any container into which it is poured. Like a solid, a liquid's volume is slightly affected by changes in temperature.

A **gas**, at a given temperature, has neither a definite shape nor a definite volume. It will take up the shape of any container into which it is placed and will spread out evenly within it. Unlike solids and liquids, the volumes of gases are affected greatly by changes in temperature.

Liquids and gases, unlike solids, are compressible. This means that their volume can be reduced by the application of pressure. Gases are much more compressible than liquids.

▲ **Figure 1.2** Without expansion gaps between the rails, the track would buckle in hot weather

1.2 The kinetic particle theory of matter

The **kinetic particle theory** helps to explain the way that matter behaves. It is based on the idea that all matter is made up of tiny particles. This theory explains the physical properties of matter in terms of the movement of the particles from which it is made.

The main points of the theory are:

» All matter is made up of tiny, moving particles, invisible to your eye. Different substances have different types of particles (atoms, molecules or ions) of varying sizes.
» The particles move all the time. The higher the temperature, the faster they move on average.
» Heavier particles move more slowly than lighter ones at a given temperature.

The kinetic particle theory can be used as a scientific model to explain how the arrangement of particles relates to the properties of the three states of matter.

Explaining the states of matter

In a solid, the particles attract one another. There are attractive forces between the particles which hold them close together. The particles have little freedom of movement and can only vibrate about a fixed position. They are arranged in a regular manner, which explains why many solids form **crystals**.

It is possible to model such crystals by using spheres to represent the particles. For example, Figure 1.3a shows spheres built up in a regular way to represent the structure of a chrome alum crystal. The shape is very similar to that of a part of an actual chrome alum crystal (Figure 1.3b).

a A model of a chrome alum crystal

b An actual chrome alum crystal

▲ **Figure 1.3**

Studies using X-ray crystallography (Figure 1.4) have confirmed how particles are arranged in crystal structures. When crystals of a pure substance form under a given set of conditions, the particles are always arranged (or packed) in the same way. However, the particles may be packed in different ways in crystals of different substances. For example, common salt (sodium chloride) has its particles arranged to give cubic crystals as shown in Figure 1.5.

▲ **Figure 1.4** A modern X-ray crystallography instrument used for studying crystal structure

▲ **Figure 1.5** Sodium chloride crystals

In a liquid, the particles are still close together but they move around in a random way and often collide with one another. The forces of attraction between the particles in a liquid are weaker than those in a solid. Particles in the liquid form of a substance have more energy on average than the particles in the solid form of the same substance.

In a gas, the particles are relatively far apart. They are free to move anywhere within the container in which they are held. They move randomly at very high velocities, much more rapidly than those in a liquid. They collide with each other, but less often than in a liquid, and they also collide with the walls of the container. They exert virtually no forces of attraction on each other because they are relatively far apart. Such forces, however, are very significant. If they did not exist, we could not have solids or liquids (see Changes of state, p. 4).

The arrangement of particles in solids, liquids and gases is shown in Figure 1.6.

Solid
Particles only vibrate about fixed positions. Regular structure.

Liquid
Particles have some freedom and can move around each other. Collide often.

Gas
Particles move freely and at random in all the space available. Collide less often than in liquid.

▲ **Figure 1.6** The arrangement of particles in solids, liquids and gases

1.3 Changes of state

The kinetic particle theory model can be used to explain how a substance changes from one state to another. If a solid is heated, the particles vibrate faster as they gain energy. This makes them 'push' their neighbouring particles further away. This causes an increase in the volume of the solid, such that the solid expands, and we can say that expansion has taken place.

Eventually, the heat energy causes the forces of attraction to weaken. The regular pattern of the structure breaks down, and the particles can now move around each other. The solid has melted. The temperature at which this takes place is called the **melting point** of the substance. The temperature of a melting pure solid will not rise until it has all melted. When the substance has become a liquid, there are still very significant forces of attraction between the particles, which is why the substance is a liquid and not a gas.

Solids which have high melting points have stronger forces of attraction between their particles than those which have low melting points. A list of some substances with their corresponding melting and boiling points is shown in Table 1.1.

▼ **Table 1.1** Melting points and boiling points of substances

Substance	Melting point/°C	Boiling point/°C
Aluminium	661	2467
Ethanol	−117	79
Magnesium oxide	827	3627
Mercury	−30	357
Methane	−182	−164
Oxygen	−218	−183
Sodium chloride	801	1413
Sulfur	113	445
Water	0	100

If a liquid is heated, the average energy of the particles increases and the particles will move around even faster. Some particles at the surface of the liquid have enough energy to overcome the forces of attraction between themselves and the other particles in the liquid and they escape to form a gas. The liquid begins to **evaporate** as a gas is formed.

Eventually, a temperature is reached at which the particles are trying to escape from the liquid so quickly that bubbles of gas actually start to form inside the liquid. This temperature is called the **boiling point** of the substance. At the boiling point, the pressure of the gas created above the liquid equals that of the air, which is **atmospheric pressure**.

Liquids with high boiling points have stronger forces between their particles than liquids with low boiling points.

When a gas is cooled, the average energy of the particles decreases and the particles move closer together. The forces of attraction between the particles now become significant and cause the gas to **condense** into a liquid. When a liquid is cooled, it freezes to form a solid. Energy is released in each of these changes.

Changes of state are examples of physical changes. Whenever a physical change of state occurs, the temperature remains constant during the change. During a physical change, no new substance is formed.

Heating and cooling curves

The graph shown in Figure 1.7 was drawn by plotting the temperature of water as it was heated steadily from −15°C to 110°C. You can see from the curve that changes of state have taken place. When the temperature was first measured, only ice was present. After a short time, the curve flattens showing that even though heat energy is being put in, the temperature remains constant.

▲ **Figure 1.7** Graph of temperature against time for the change from ice at −15°C to water to steam

Practical skills

Changes of state

For safe experiments/demonstrations which are related to this chapter, please refer to the *Cambridge IGCSE Chemistry Practical Skills Workbook*, which is also part of this series.

Safety

- Eye protection must be worn.
- Take care when handling and using hot water.

The apparatus on the right was set up to obtain a cooling curve for stearic acid. The stearic acid was placed into a boiling tube which was then placed in a beaker of water that was heated to 80°C, which is above the melting point of stearic acid.

The boiling tube was then removed from the beaker and the temperature of the stearic acid was recorded every minute for 12 minutes using the thermometer to stir the stearic acid while it was a liquid.

1 Why was it important to remove the boiling tube with the stearic acid from the water?
2 Why was the stearic acid stirred with the thermometer?
3 Why were temperature readings taken every minute for 12 minutes?

The following data was obtained from the experiment:

Time/mins	0	1	2	3	4	5	6	7	8	9	10	11	12
Temperature/°C	79	76	73	70	69	69	69	69	69	67	64	62	60

4 Draw and label axes for plotting this data.
5 Plot the points and draw a line of best fit.

6 a At what temperature did the stearic acid begin to change state?
  b How could you tell this from your graph?
  c Explain what is happening at this temperature.

In ice, the particles of water are close together and are attracted to one another. For ice to melt, the particles must obtain sufficient energy to overcome the forces of attraction between the water particles, so that relative movement can take place. The heat energy is being used to overcome these forces.

The temperature will begin to rise again only after all the ice has melted. Generally, the heating curve for a pure solid always stops rising at its melting point and gives rise to a sharp melting point. A sharp melting point therefore indicates that it is a pure sample. The addition or presence of impurities lowers the melting point.

You can find the melting point of a substance using the apparatus shown in Figure 1.8. The addition or presence of impurities lowers the melting point. A mixture of substances also has a lower melting point than a pure substance, and the melting point will be over a range of temperatures and not sharp.

▲ **Figure 1.8** If a substance, such as the solid in the melting point tube, is heated slowly, this apparatus can be used to find the melting point of the substance

In the same way, if you want to boil a liquid, such as water, you have to give it some extra energy. This can be seen on the graph in Figure 1.7, where the curve levels out at 100°C – the boiling point of water.

Solids and liquids can be identified from their characteristic melting and boiling points.

The reverse processes of condensing and freezing occur when a substance is cooled. Energy is given out when the gas condenses to the liquid and the liquid freezes to give the solid.

1.4 The effects of temperature and pressure on the volume of a gas

▲ **Figure 1.9** Temperature changes the volume of the air in a balloon. Higher temperatures increase the volume of the balloon and cold temperatures reduce its volume.

What do you think has caused the difference between the balloons in Figure 1.9? The pressure inside a balloon is caused by the gas particles striking the inside surface of the balloon. At a higher temperature, there is an increased pressure inside the balloon (Figure 1.10). This is due to the gas particles having more energy and therefore moving around faster, which results in the particles striking the inside surface of the balloon more frequently, which leads to an increase in pressure.

▲ **Figure 1.10** The gas particles striking the surface create the pressure

Since the balloon is made from an elastic material, the increased pressure causes the balloon to stretch and the volume increases. An increase in volume of a gas with increased temperature is a property of all gases. French scientist J.A.C. Charles made an observation like this in 1781 and concluded that when the temperature of a gas increased, the volume also increased at a fixed pressure. We can extend this idea to suggest that changing the pressure of a fixed volume of a gas must have an effect on the temperature of the gas. If you have ever used a bicycle pump to blow up a bicycle tyre then you may have felt the pump get hotter the more you used it. As you use the pump you increase pressure on the air in the pump. Such an increase in pressure causes the gas molecules to move closer together so the molecules to collide more frequently and more frictional forces come into play, which causes the temperature to rise. In addition, as the molecules are forced closer to one another, intermolecular bonds form, again increasing the temperature of the gas. As the temperature of the gas increases, this also causes the molecules to move faster, causing even more collisions.

Test yourself

2 Why do gases expand more than solids for the same increase in temperature?
3 Ice on a car windscreen will disappear as you drive along, even without the heater on. Explain why this happens.
4 When salt is placed on ice, the ice melts. Explain why this happens.
5 Draw and label a graph of water at 100°C being allowed to cool to –5°C.

1.5 Diffusion

When you go through the door of a restaurant, you can often smell the food being cooked. For this to happen, gas particles must be leaving the pans the food is being cooked in and be spreading out through the air in the restaurant. This spreading out of a gas is called **diffusion** and it takes place in a haphazard and random way.

All gases diffuse to fill the space available. Figure 1.11 shows two gas jars on top of each other. Liquid bromine has been placed in the bottom gas jar (left photo) and then left for a day (right photo). The brown-red fumes are gaseous bromine that has spread evenly throughout both the gas jars from the liquid present in the lower gas jar.

▲ **Figure 1.11** After 24 hours the bromine fumes have diffused throughout both gas jars

Diffusion can be explained by the **kinetic particle theory**. This theory states that all matter is made up of many small particles which are constantly moving. In a solid, as we have seen, the particles simply vibrate about a fixed point. However, in a gas, the particles move randomly past one another, colliding with each other.

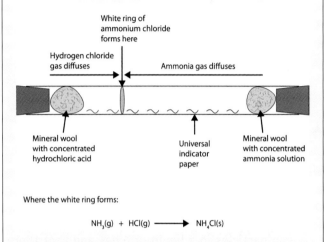

Where the white ring forms:

$$NH_3(g) + HCl(g) \longrightarrow NH_4Cl(s)$$

▲ **Figure 1.12** Hydrochloric acid (left) and ammonia (right) diffuse at different rates

Gases diffuse at different rates. If one piece of cotton wool is soaked in concentrated ammonia solution and another is soaked in concentrated hydrochloric acid and these are put at opposite ends of a dry glass tube, then after a few minutes a white cloud of ammonium chloride appears. Figure 1.12 shows the position at which the two gases meet and react. The white cloud forms in the position shown because the ammonia particles are lighter; they have a smaller relative molecular mass (Chapter 4, p. 54) than the hydrogen chloride particles (released from the hydrochloric acid) and so move faster, such that the gas diffuses more quickly. (See *Chemistry Practical Skills Workbook* for more detail of this experiment.) This experiment is a teacher demonstration only, which must be carried out in a fume cupboard. If considering carrying out this practical, teachers should refer to the *Practical Skills Workbook* for full guidance and safety notes.

Diffusion also takes place in liquids (Figure 1.13) but it is a much slower process than in gases. This is because the particles of a liquid move much more slowly.

▲ **Figure 1.13** Diffusion of green food colouring can take days to reach the stage shown on the right

When diffusion takes place between a liquid and a gas, it is known as intimate mixing. Kinetic particle theory can be used to explain this process. It states that collisions are taking place randomly between particles in a liquid or a gas and that there is sufficient space between the particles of one substance for the particles of the other substance to move into.

Test yourself

6 When a jar of coffee is opened, people can often smell it from anywhere in the room. Use the kinetic particle theory to explain how this happens.
7 Describe, with the aid of diagrams, the diffusion of a drop of green food colouring added to the bottom of a beaker.
8 Explain why diffusion is faster in gases than in liquids.

9 Explain why a gas with a low relative molecular mass can diffuse faster than a gas with a high relative molecular mass at the same temperature.

Revision checklist

After studying Chapter 1 you should be able to:
✔ State the three states of matter and describe the structure arrangement of the particles in each.
✔ Explain the properties of the three states of matter using ideas about the separation and movement of particles.
✔ Name the changes of state and describe what happens to the particles in a substance when they occur.

✔ Explain what is happening when a substance changes state.
✔ Describe what happens to a given amount of gas when temperature and/or pressure decreases and increases.
✔ Use the kinetic particle theory to explain the effects of pressure and temperature on the volume of a gas.
✔ Describe the process of diffusion and explain why gases diffuse.
✔ Describe and explain the rate of diffusion of a gas in terms of its relative molecular mass.

Exam-style questions

1 a Sketch diagrams to show the arrangement of particles in:
  i solid oxygen [1]
  ii liquid oxygen [1]
  iii oxygen gas. [1]
 b Describe how the particles move in these three states of matter. [3]

 c Explain, using the kinetic particle theory, what happens to the particles in oxygen as it is cooled down. [3]

2 Explain the meaning of each of the following terms. In your answer include an example to help with your explanation.
 a expansion [2]
 b contraction [2]
 c physical change [2]
 d diffusion [2]
 e random motion [2]
3 a Explain why solids do not diffuse. [2]
 b Give two examples of diffusion of gases and liquids found in your house. [2]
4 Explain the following, using the ideas you have learned about the kinetic particle theory:
 a When you take a block of butter out of the fridge, it is quite hard. However, after 15 minutes it is soft enough to spread. [2]
 b When you come home from school and open the door, you can smell food being cooked. [2]
 c A football is blown up until it is hard on a hot summer's day. In the evening the football feels softer. [2]
 d When a person wearing perfume enters a room, it takes several minutes for the smell to reach the back of the room. [2]
5 Some green food colouring was carefully added to the bottom of a beaker of water using a syringe. The beaker was then covered and left for several days.

Beaker

Water

Green food colouring

a Describe what you would observe after:
  i a few hours [1]
  ii several days. [1]

 b Explain your answer to Question 5a using your ideas of the kinetic particle theory. [2]

 c State the physical process that takes place in this experiment. [1]

6 The apparatus shown below was set up.

Stopper   White cloud   Glass tube

Cotton wool soaked in concentrated hydrochloric acid

Cotton wool soaked in concentrated ammonia solution

When this apparatus is used, the following things are observed. Explain why each of these is observed.
 a A white cloud is formed. [1]
 b It took a few minutes before the white cloud formed. [1]
 c The white cloud formed further from the cotton wool soaked in ammonia than that soaked in hydrochloric acid. [2]
 d Cooling the concentrated ammonia and hydrochloric acid before carrying out the experiment increased the time taken for the white cloud to form. [1]

2 Atoms, elements and compounds

FOCUS POINTS

★ How are elements, molecules, ions, compounds and mixtures different from each other?
★ How do the properties of the particles in an atom lead to an atom's structure?
★ What do oxidation and reduction mean?
★ What is an isotope?

In Chapter 1, you saw that all matter is made up of particles. In this chapter you will look closely at these particles and see that they are made up of atoms. Atoms are the smallest part of elements. An element is made up of one type of atom and can be either a metal or a non-metal. Metals and non-metals have different properties.

You will look at how atoms of different elements can combine to form substances called compounds, and how this combining occurs in a chemical reaction. By the end of the chapter you should be able to write a simple word or symbol equation to represent these reactions.

You will see that although atoms are the smallest part of an element that shares the chemical properties of that element, they are made from even smaller particles. By learning about the properties and behaviour of these smaller particles (electrons, protons and neutrons), you will be able to see how they affect the chemical properties of elements and compounds.

The universe is made up of a very large number of substances (Figure 2.1), and our own part of the universe is no exception. When we examine this vast array of substances more closely, it is found that they are made up of some basic substances which were given the name **elements** in 1661 by Robert Boyle.

▲ **Figure 2.1** Structures in the universe, such as stars, planets and meteorites, are made of millions of substances. These are made up mainly from just 91 elements, all of which occur naturally on the Earth

In 1803, John Dalton suggested that each element was composed of its own kind of particles, which he called **atoms**. Atoms are much too small to be seen. We now know that about 20×10^6 of them would stretch over a length of only 1 cm.

2.1 Elements

As well as not being able to be broken down into a simpler substance, each element is made up of only one kind of atom. The word atom comes from the Greek word *atomos* meaning 'unsplittable'. For example, aluminium is an element which is made up of only aluminium atoms. It is not possible to obtain a simpler substance chemically from the aluminium atoms. You can only combine it with other elements to make more complex substances, such as aluminium oxide, aluminium nitrate or aluminium sulfate.

One hundred and eighteen elements have now been identified. Twenty of these do not occur in nature and have been made artificially by scientists. They include elements such as curium and flerovium. Ninety-eight of the elements occur naturally and range from some very reactive gases, such as fluorine and chlorine, to gold and platinum, which are unreactive elements. A physical property is any characteristic of a substance that we can measure. The elements have different properties that we can measure, and we can then classify them according to those properties.

All elements can be classified according to their various properties. A simple way to do this is to classify them as **metals** or **non-metals** (Figures 2.2 and 2.3). Table 2.1 shows the physical property data for some common metallic and non-metallic elements. You will notice from Table 2.1 that many metals have high densities, high melting points and high boiling points, and that most non-metals have low densities, low melting points and low boiling points. Table 2.2 summarises the different properties of metals and non-metals.

a   Gold is very decorative

b   Aluminium has many uses in the aerospace industry

c   These coins contain nickel

▲ **Figure 2.2** Some metals

▼ **Table 2.1** Physical data for some metallic and non-metallic elements at room temperature and pressure

Element	Metal or non-metal	Density/ g cm⁻³	Melting point/°C	Boiling point/°C
Aluminium	Metal	2.70	660	2580
Copper	Metal	8.92	1083	2567
Gold	Metal	19.29	1065	2807
Iron	Metal	7.87	1535	2750
Lead	Metal	11.34	328	1740
Magnesium	Metal	1.74	649	1107
Nickel	Metal	8.90	1453	2732
Silver	Metal	10.50	962	2212
Zinc	Metal	7.14	420	907
Carbon	Non-metal	2.25	Sublimes at 3642	
Hydrogen	Non-metal	0.07[a]	−259	−253
Nitrogen	Non-metal	0.88[b]	−210	−196
Oxygen	Non-metal	1.15[c]	−218	−183
Sulfur	Non-metal	2.07	113	445

Source: Earl B., Wilford L.D.R. Chemistry data book. Nelson Blackie, 1991 a: at −254°C; b: at −197°C; c: at −184°C.

The elements also have chemical properties, which are characteristics or behaviours that may be observed when the substance undergoes a chemical change or reaction. A discussion of the chemical properties of some metals and non-metals is given in Chapters 9 and 10.

▼ **Table 2.2** How the properties of metals and non-metals compare

Property	Metal	Non-metal
Physical state at room temperature	Usually solid (occasionally liquid)	Solid, liquid or gas
Malleability	Good	Poor – usually soft or brittle
Ductility	Good	
Appearance (solids)	Shiny (lustrous)	Dull
Melting point	Usually high	Usually low
Boiling point	Usually high	Usually low
Density	Usually high	Usually low
Conductivity (thermal and electrical)	Good	Very poor

c   Neon is used in advertising signs

▲ **Figure 2.3** Some non-metals

Atoms – the smallest particles

Everything is made up of billions of extremely small atoms. The smallest atom is hydrogen, and we represent each hydrogen atom as a sphere having a diameter of 0.000 000 07 mm (or 7×10^{-8} mm) (Table 2.3). Atoms of different elements have different diameters as well as different masses.

▼ **Table 2.3** Sizes of atoms

Atom	Diameter of atom/mm	Masses/g
Hydrogen	7×10^{-8}	1.67×10^{-24}
Oxygen	12×10^{-8}	2.66×10^{-23}
Sulfur	20.8×10^{-8}	5.32×10^{-23}

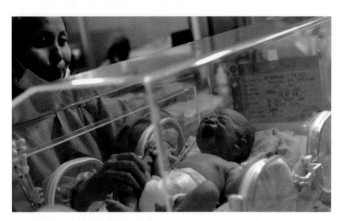

a   A premature baby needs oxygen

Chemists use shorthand symbols to label the elements and their atoms. The symbol consists of one, two or three letters, the first of which is always a capital. The initial letter of the element's name is often used and, where several elements have the same initial letter, another letter from the name is added. For example, **C** is used for **carbon**, **Ca** for **calcium** and **Cl** for **chlorine**. Some symbols seem to have no relationship to the name of the element, for example, **Na** for **sodium** and **Pb** for **lead**. These symbols come from their Latin names: **na**trium for sodium and **p**lum**b**um for lead. A list of some common elements and their symbols is given in Table 2.4.

b   Artists often use charcoal (carbon) to produce an initial sketch

▼ **Table 2.4** Some common elements and their symbols. The Latin names of some of the elements are given in brackets

Element	Symbol	Physical state at room temperature and pressure
Aluminium	Al	Solid
Argon	Ar	Gas
Barium	Ba	Solid
Boron	B	Solid
Bromine	Br	Liquid
Calcium	Ca	Solid
Carbon	C	Solid
Chlorine	Cl	Gas
Chromium	Cr	Solid
Copper (Cuprum)	Cu	Solid
Fluorine	F	Gas
Germanium	Ge	Solid
Gold (Aurum)	Au	Solid
Helium	He	Gas
Hydrogen	H	Gas
Iodine	I	Solid
Iron (Ferrum)	Fe	Solid
Lead (Plumbum)	Pb	Solid
Magnesium	Mg	Solid
Mercury (Hydragyrum)	Hg	Liquid
Neon	Ne	Gas
Nitrogen	N	Gas
Oxygen	O	Gas
Phosphorus	P	Solid
Potassium (Kalium)	K	Solid
Silicon	Si	Solid
Silver (Argentum)	Ag	Solid
Sodium (Natrium)	Na	Solid
Sulfur	S	Solid
Tin (Stannum)	Sn	Solid
Zinc	Zn	Solid

Molecules

The atoms of some elements are joined together in small groups. These small groups of atoms are called **molecules**. The atoms of some elements are always joined in pairs, for example, hydrogen, oxygen, nitrogen, fluorine, chlorine, bromine and iodine. They are known as **diatomic** molecules. In chemical shorthand the molecule of chlorine shown in Figure 2.4 is written as Cl_2. The atoms of some other elements, such as phosphorus and sulfur, join in larger numbers, four and eight respectively, which we write as P_4 and S_8.

The complete list of the elements with their corresponding symbols is shown in the **Periodic Table** on p. 135.

The gaseous elements helium, neon, argon, krypton, xenon and radon (which are all gases at 0°C at sea level and atmospheric pressure) are composed of separate, individual atoms. When an element exists as separate atoms, then the molecules are said to be **monatomic**. In chemical shorthand these monatomic molecules are written as He, Ne, Ar, Kr, Xe and Rn respectively.

Cl——Cl

a   Represented by a letter-and-stick model

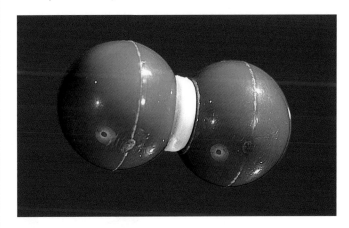

b   Represented by a space-filling model

▲ **Figure 2.4** A chlorine molecule

Molecules are not always formed by atoms of the same type joining together as elemental molecules. Most molecules consist of atoms of different elements, for example, water exists as molecules containing oxygen and hydrogen atoms. We will learn more about these in the next section.

> ### Test yourself
>
> 4 How many atoms of hydrogen would have to be placed side by side along the edge of your ruler to fill just one of the 1 mm divisions?
> 5 How would you use chemical shorthand to write a representation of the molecules of iodine and fluorine?
> 6 Using the Periodic Table on p. 135, write down the symbols for each of these elements and give their physical state at room temperature.
>   a chromium
>   b krypton
>   c osmium

2.2 Compounds

Compounds are pure substances which are formed when two or more elements chemically combine together. A **pure substance** is a material that has a constant composition (is homogeneous) and has consistent properties throughout. Water is a simple compound formed from the elements hydrogen and oxygen (Figure 2.5). This combining of the elements can be represented by a word equation:

hydrogen + oxygen   →   water

| Hydrogen a pure element | Oxygen a pure element | Hydrogen and oxygen mixed together | Water a pure compound formed from hydrogen burning in oxygen |

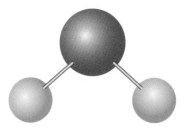

b A model of water showing 2 H atoms and one O atom. Models such as this can be built to show what a compound looks like

▲ **Figure 2.5**

Water molecules contain two atoms of hydrogen and one atom of oxygen, and water has the **chemical formula** H_2O. If there is only one atom of an element in the molecule, no number is required in the formula, only its symbol, as in the case of oxygen in the water molecule H_2O.

> ### Test yourself
>
> 7 What is the formula for the molecule shown in the diagram which contain carbon (black sphere) and hydrogen (white spheres)?
>
>

Elements other than hydrogen will also react with oxygen to form compounds called oxides. For example, magnesium reacts violently with oxygen gas to form the white powder magnesium oxide (Figure 2.6). This reaction is accompanied by a release of energy as new chemical bonds are formed.

a The element hydrogen reacts with the element oxygen to produce the compound water

Any chemical process that involves reduction and oxidation is known as a **redox** reaction. For example, to extract iron from iron(III) oxide, the oxygen has to be removed. The reduction of iron(III) oxide can be done in a **blast furnace** using carbon monoxide. The iron(III) oxide loses oxygen to the carbon monoxide and is reduced to iron.

> Carbon monoxide is the **reducing agent**. A reducing agent is a substance that reduces another substance during a redox reaction. In the reaction, carbon monoxide is oxidised to carbon dioxide by the iron(III) oxide. In this process, the iron(III) oxide is the **oxidising agent**. An oxidising agent is a substance which oxidises another substance during a redox reaction.

We can write the redox reaction as:

iron(III) oxide + carbon monoxide → iron + carbon dioxide

▲ **Figure 2.6** Magnesium burns brightly in oxygen to produce magnesium oxide

When a new substance is formed during a chemical reaction, a **chemical change** has taken place.

magnesium + oxygen → magnesium oxide

When substances such as hydrogen and magnesium combine with oxygen in this way, they are said to have been oxidised, and this process is known as **oxidation**.

Reduction is the opposite of oxidation. In this process oxygen is removed rather than added.

Key definitions

Oxidation is gain of oxygen.

Reduction is loss of oxygen.

Test yourself

8   Zinc is extracted from its ore zinc blende in a furnace by a redox reaction. What does the term 'redox reaction' mean?

9   Identify the oxidising and reducing agents in each of the following reactions:
   a   copper(II) oxide + hydrogen → copper + water
   b   tin(II) oxide + carbon → tin + carbon dioxide
   c $PbO(s) + H_2(g) → Pb(s) + H_2O(l)$.

For a further discussion of oxidation and reduction see Chapter 3 (p. 31) and Chapter 5 (p. 71).

Key definitions

Redox reactions involve simultaneous oxidation and reduction.

An **oxidising agent** is a substance that oxidises another substance and is itself reduced.

A **reducing agent** is a substance that reduces another substance and is itself oxidised.

Practical skills

Heating copper

For safe experiments/demonstrations which are related to this chapter, please refer to the *Cambridge IGCSE Chemistry Practical Skills Workbook*, which is also part of this series.

Safety

- Eye protection must be worn.
- Take care when handling hot apparatus.
- Handle the copper with tongs or tweezers, not your fingers.

A student wants to find out what happens when copper is heated in air. In order to do this, they carried out the following experiment and recorded their results.

- First, they found the mass of an empty crucible (a suitably prepared beer-bottle top (metal) is an alternative to a porcelain crucible).
- They added a piece of copper to the crucible and found the mass again.
- They then heated the crucible strongly for approximately two minutes.
- After they allowed it to cool, they then found the mass after heating.

Mass of crucible = 12.90 g
Mass of crucible + copper = 14.18 g

Mass of copper = _____ g
Mass of crucible + contents after heating = 14.30 g

Colour of contents after heating = black

1 Draw a labelled diagram of the experimental set-up used in this experiment.
2 Calculate the change in mass that has taken place during the heating.
3 Explain what has caused the change in mass.
4 What is the black substance left on the copper after heating?
5 Write a word and balanced chemical equation to show the process that has taken place.
6 a How could you modify the experiment to ensure there was no loss of substance taking place during the heating process?
   b What are the other possible sources of error?
7 Predict what would happen, in terms of mass change and colour change, if calcium were heated in air in the same way as the copper.

Formulae

The formula of a compound is made up from the symbols of the elements that make up the compound and numbers that show the ratio of the different atoms the compound is made from. Carbon dioxide has the formula CO_2, which tells you that it contains one carbon atom for every two oxygen atoms. The 2 in the formula indicates that there are two oxygen atoms present in each molecule of carbon dioxide.

? Worked example

Write the ratio of atoms in sodium sulfate – Na_2SO_4.

Substance	Formula	Ratio of atoms
Sodium sulfate	Na_2SO_4	Na : S : O 2 : 1 : 4

► Test yourself

10 Write down the ratio of the atoms present in the formula for each of the compounds shown in Table 2.5.

▼ **Table 2.5** Names and formulae of some common compounds

Compound	Formula
Ammonia	NH_3
Calcium hydroxide	$Ca(OH)_2$
Carbon dioxide	CO_2
Copper sulfate	$CuSO_4$
Ethanol (alcohol)	C_2H_5OH
Glucose	$C_6H_{12}O_6$
Hydrochloric acid	HCl
Nitric acid	HNO_3
Sodium carbonate	Na_2CO_3
Sodium hydroxide	NaOH
Sulfuric acid	H_2SO_4

The ratio of atoms within a chemical compound is usually constant. Compounds are made up of fixed proportions of elements: they have a fixed composition. Chemists call this the **Law of constant composition.**

For further discussion of formulae, see p. 35.

Balancing chemical equations

Word equations are a useful way of representing chemical reactions, but a better and more useful way of seeing what happens during a chemical reaction is to produce a balanced chemical equation. This type of equation gives the formulae of the substances that are reacting, the reactants, and the new substances formed during the chemical reaction, the products, as well as showing the relative numbers of each of the particles involved.

Balanced equations often include symbols that show the physical state of each of the reactants and products:

(s) = solid
(l) = liquid
(g) = gas
(aq) = aqueous (water) solution

We can use the reaction between iron and sulfur as an example. The word equation to represent this reaction is:

$$\text{iron} + \text{sulfur} \xrightarrow{\text{heat}} \text{iron(II) sulfide}$$

When we replace the words with symbols for the reactants and the products, and include their physical state symbols, we get:

$$Fe(s) + S(s) \xrightarrow{\text{heat}} FeS(s)$$

Since there is the same number of each type of atom on both sides of the equation, this is a balanced chemical equation.

❓ Worked example

Write, for the reaction between magnesium and oxygen producing magnesium oxide:
a   the word equation
b   the balanced chemical equation.

a   The word equation is:

$$\text{magnesium} + \text{oxygen} \xrightarrow{\text{heat}} \text{magnesium oxide}$$

b   When we replace the words with symbols for the reactants and the products and include their physical state symbols, it is important to remember that oxygen is a diatomic molecule:

$$Mg(s) + O_2(g) \xrightarrow{\text{heat}} MgO(s)$$

In the equation there are two oxygen atoms on the left-hand side (O_2) but only one on the right (MgO). We cannot change the formula of magnesium oxide, so to produce the necessary two oxygen atoms on the right-hand side, we will need 2MgO – this means 2 × MgO formula units. The equation now becomes:

$$Mg(s) + O_2(g) \xrightarrow{\text{heat}} 2MgO(s)$$

There are now two atoms of magnesium on the right-hand side and only one on the left. To balance the equation, we place a 2 in front of the magnesium, and obtain the following balanced chemical equation:

$$2Mg(s) + O_2(g) \xrightarrow{\text{heat}} 2MgO(s)$$

This balanced chemical equation now shows us that two atoms of magnesium react with one molecule of oxygen gas when heated to produce two units of magnesium oxide.

▶ Test yourself

11 Write the word and balanced chemical equations for the reactions which take place between:
   a   calcium and oxygen
   b   copper and oxygen.

Going further

Instrumental techniques

Elements and compounds can be detected and identified by a variety of instrumental methods. Scientists have developed instrumental techniques that allow us to probe and discover which elements are present in a substance as well as how the atoms are arranged within the substance.

Many of the instrumental methods that have been developed are quite sophisticated. Some methods are suited to identifying elements. For example, atomic absorption spectroscopy allows the element to be identified and also allows the quantity of the element that is present to be found.

Some methods are particularly suited to the identification of compounds. For example, infrared spectroscopy is used to identify compounds by showing the presence of particular groupings of atoms.

Infrared spectroscopy is used by the pharmaceutical industry to identify and discriminate between drugs that are similar in structure, for example, penicillin-type drugs. Used both with organic and inorganic molecules, this method assumes that each compound has a unique infrared spectrum. Samples can be solid, liquid or gas and are usually tiny. However, Ne, He, O_2, N_2 or H_2 cannot be used.

This method is also used to monitor environmental pollution and has biological uses in monitoring tissue physiology including oxygenation, respiratory status and blood-flow damage.

Forensic scientists make use of both these techniques because they are very accurate, but they only require tiny amounts of sample – often only small amounts of sample are found at crime scenes. Other techniques utilised are nuclear magnetic resonance spectroscopy and ultraviolet/visible spectroscopy.

2.3 Mixtures

Many everyday things are not pure substances: they are **mixtures**. A mixture contains more than one substance, which could be elements and/or compounds. Examples of common mixtures are:

» sea water (Figure 2.7)
» air, which is a mixture of elements such as oxygen, nitrogen and neon, and compounds such as carbon dioxide (see Chapter 11, p. 178)
» alloys such as brass, which is a mixture of copper and zinc (for a further discussion of alloys, see Chapter 10, p. 165).

▲ **Figure 2.7** Sea water is a common mixture. It is a water solution of substances such as sodium chloride as well as gases such as oxygen and carbon dioxide

The difference between mixtures and compounds

There are differences between compounds and mixtures, which can be seen by looking at the reaction between iron filings and sulfur. A mixture of iron filings (powdered iron) and sulfur (Figure 2.8 bottom right), looks different from either of the individual elements (Figure 2.8 top). This mixture has the properties of both iron and sulfur; for example, a magnet can be used to separate the iron filings from the sulfur (Figure 2.9).

▲ **Figure 2.8** The elements sulfur and iron at the top of the photograph, and (below) black iron(II) sulfide on the left and a mixture of the two elements on the right

Substances in a mixture have not undergone a chemical reaction and it is possible to separate them, provided that there is a suitable difference in their physical properties. If the mixture of iron and sulfur is heated, a chemical reaction occurs and a new substance is formed. The product of the reaction is iron(II) sulfide (Figure 2.8 bottom left), and the word equation for this reaction is:

$$\text{iron} + \text{sulfur} \xrightarrow{\text{heat}} \text{iron(II) sulfide}$$

▲ **Figure 2.9** A magnet will separate the iron from the mixture

During the reaction, heat energy is given out as new chemical bonds are formed. This is called an **exothermic reaction** and accompanies a chemical change (Chapter 6, p. 92). The iron(II) sulfide formed has very different properties to the mixture of iron and sulfur (Table 2.6); for example, iron(II) sulfide would not be attracted towards a magnet. Some chemical reactions take in heat during the

reaction, which is called an **endothermic reaction** (Chapter 6, p. 92). You will learn more about the different types of reactions in Chapter 6.

▼ **Table 2.6** Different properties of iron, sulfur, an iron/sulfur mixture and iron(II) sulfide

Substance	Appearance	Effect of a magnet	Effect of dilute hydrochloric acid
Iron	Dark grey powder	Attracted to it	Very little action when cold. When warm, a gas is produced with a lot of bubbling (effervescence)
Sulfur	Yellow powder	None	No effect when hot or cold
Iron/sulfur mixture	Dirty yellow powder	Iron powder attracted to it	Iron powder reacts as above
Iron(II) sulfide	Black solid	No effect	A foul-smelling gas is produced with some effervescence

In iron(II) sulfide, FeS, one atom of iron has combined with one atom of sulfur. In a mixture of iron and sulfur, no such ratio exists as the atoms have not chemically combined. Table 2.7 compares mixtures and compounds. Some common mixtures are discussed in Chapter 10 (p. 165) and Chapter 11 (p. 177).

▼ **Table 2.7** The major differences between mixtures and compounds

Mixture	Compound
It contains two or more substances.	It is a single substance.
The composition can vary.	The composition is always the same.
No chemical change takes place when a mixture is formed.	When the new substance is formed it involves chemical change.
The properties are those of the individual elements/compounds.	The properties are very different to those of the component elements.
The components may be separated quite easily by physical means.	The components can only be separated by one or more chemical reactions.

> # Test yourself
>
> 12 Make a list of some other common mixtures and then use your research to find out and state what they are mixtures of.
>
> 13 Which of the following are not mixtures: milk, tin, sulfur, cough linctus, brass, gold?

Going further

Other types of mixtures

There are mixtures which are formed by mixing two substances (or phases) which cannot mix. Gels, sols, foams and emulsions are all examples of just such mixtures. Look closely at the substances in Figure 2.10, which shows examples of these different types of mixture.

a

b

c

d

a This jelly is an example of a 'gel'
b Emulsion paint is an example of a 'sol'
c These foams have been formed by trapping bubbles of gas in liquids
d Emulsions are formed by mixing immiscible liquids

▲ **Figure 2.10**

Composite materials are those that combine the properties of two constituents in order to get the exact properties needed for a particular job. Glass-reinforced fibre is an example of a composite material combining the properties of two different materials. It is made by embedding short fibres of glass in a matrix of plastic.

The glass fibres give the plastic extra strength so that it does not break when it is bent or moulded into shape. The finished material has the strength and flexibility of the glass fibres as well as the lightness of plastic (Figure 2.11).

a Glass-reinforced plastic (GRP) consists of glass fibres (rod shapes) embedded in plastic, in this case polyester

b The glass-reinforced plastic used to make boats like this is a composite material

▲ **Figure 2.11**

2.4 Inside atoms

Everything you see around you is made out of tiny particles, which we call atoms (Figure 2.12). When John Dalton developed his atomic theory, over 200 years ago, he stated that the atoms of any one element were identical and that each atom was 'indivisible'. Scientists in those days also believed that atoms were solid particles, like marbles.

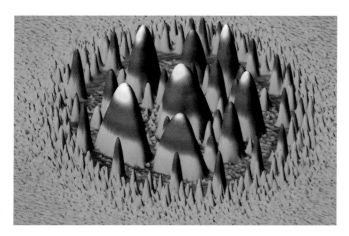

▲ **Figure 2.12** An electron micrograph of atoms taken using a very powerful microscope called an electron microscope

However, in the last hundred years or so it has been proved by great scientists, such as Niels Bohr, Albert Einstein, Henry Moseley, Joseph Thomson, Ernest Rutherford and James Chadwick, that atoms are in fact made up of even smaller 'sub-atomic' particles. Seventy sub-atomic particles have now been discovered, and the most important of these are **electrons**, **protons** and **neutrons**.

These three sub-atomic particles are found in distinct and separate regions of the atom. The protons and neutrons are found in the centre of the atom, which is called the **nucleus**. Neutrons have no charge and protons are positively charged. The nucleus occupies only a very small volume of the atom and is very dense.

The rest of the atom surrounding the nucleus is where electrons are found. Electrons are negatively charged and move around the nucleus very quickly at specific distances from the nucleus in **electron shells** or **energy levels**. The electrons are held in the shells within the atom by an **electrostatic force of attraction** between themselves and the positive charge of protons in the nucleus (Figure 2.13). Each shell can contain only a fixed number of electrons: the first shell can hold up to two electrons, the second shell can hold up to eight electrons, the third shell can hold up to 18, and so on.

▲ **Figure 2.13** Diagram of an atom

Nucleus, containing neutrons and protons

Electron shells

Electrons are tiny and relatively light; approximately 1837 electrons are equal in mass to the mass of one proton or one neutron. A summary of each type of particle, its mass and relative charge is shown in Table 2.8. You will notice that the masses of all these particles are measured in atomic mass units (amu). This is because they are so light that their masses cannot be measured usefully in grams.

▼ **Table 2.8** Characteristics of a proton, a neutron and an electron

Particle	Symbol	Relative mass/ amu	Relative charge
Proton	p	1	+1
Neutron	n	1	0
Electron	e	1/1837	−1

Although atoms contain electrically charged particles, the atoms themselves are electrically neutral (they have no overall electric charge). This is because atoms contain equal numbers of electrons and protons. For example, Figure 2.14 represents the atom of the non-metallic element helium. The atom of helium possesses two protons, two neutrons and two electrons. The electrical charge of the protons in the nucleus is, therefore, balanced by the opposite charge of the two electrons.

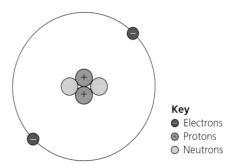

Key
● Electrons
◉ Protons
○ Neutrons

▲ **Figure 2.14** An atom of helium has two protons, two electrons and two neutrons

Proton number and mass number

The number of protons in the nucleus of an atom is called the **proton number** (or atomic number) and is given the symbol Z. The helium atom in Figure 2.14, for example, has a proton number of 2, since it has two protons in its nucleus. Each element has its own proton number and no two elements have the same proton number. For example, the element lithium has a proton number of 3 since it has three protons in its nucleus.

Neutrons and protons have a similar mass, whereas electrons possess very little mass. So the mass of any atom depends on the number of protons and neutrons in its nucleus. The total number of protons and neutrons found in the nucleus of an atom is called the **mass number** (or nucleon number) and is given the symbol A.

> **Key definitions**
>
> **Proton number** or atomic number is the number of protons in the nucleus of an atom.
>
> **Mass number** or nucleon number is the total number of protons and neutrons in the nucleus of an atom.

$$\text{mass number} = \text{proton number} + \text{number of neutrons}$$
$$(A) \qquad\qquad (Z)$$

The helium atom in Figure 2.14 has a mass number of 4, since it has two protons and two neutrons in its nucleus. If we consider the metallic element lithium, it has three protons and four neutrons in its nucleus. It therefore has a mass number of 7.

The proton number and mass number of an element are usually written in the following shorthand way:

Mass number (A) ↘ $^4_2$He ← Symbol of the element
Proton number (Z) ↗

The number of neutrons present can be calculated by rearranging the relationship between the proton number, mass number and number of neutrons to give:

$$\text{number of neutrons} = \text{mass number} - \text{proton number}$$
$$(A) \qquad\qquad (Z)$$

? Worked example

1 What is the number of neutrons in one atom of $^{24}_{12}$Mg?

number of neutrons = mass number − proton number

$12 = 24(A) - 12(Z)$

2 What is the number of neutrons in one atom of $^{207}_{82}$Pb?

number of neutrons = mass number − proton number

$125 = 207(A) - 82(Z)$

Table 2.9 shows the number of protons, neutrons and electrons in the atoms of some common elements.

▼ **Table 2.9** Number of protons, neutrons and electrons in some elements

Element	Symbol	Proton number	Number of electrons	Number of protons	Number of neutrons	Mass number
Hydrogen	H	1	1	1	0	1
Helium	He	2	2	2	2	4
Carbon	C	6	6	6	6	12
Nitrogen	N	7	7	7	7	14
Oxygen	O	8	8	8	8	16
Fluorine	F	9	9	9	10	19
Neon	Ne	10	10	10	10	20
Sodium	Na	11	11	11	12	23
Magnesium	Mg	12	12	12	12	24
Sulfur	S	16	16	16	16	32
Potassium	K	19	19	19	20	39
Calcium	Ca	20	20	20	20	40
Iron	Fe	26	26	26	30	56
Zinc	Zn	30	30	30	35	65

Ions

An **ion** is an electrically charged particle. When an atom loses one or more electrons, it is no longer electrically neutral and becomes a positively charged ion. This is called a cation.

For example, when potassium is involved in a chemical reaction, each atom loses an electron to form a positive ion, K^+.

$$_{19}K^+ \quad \frac{\begin{array}{l}19 \text{ protons} = 19+ \\ 18 \text{ electrons} = 18-\end{array}}{\text{Overall charge} = \quad 1+}$$

When an atom gains one or more electrons, it becomes a negatively charged ion. This is called an anion. For example, in some of the chemical reactions involving oxygen, each oxygen atom gains two electrons to form a negative ion, O^{2-}.

$$_{8}O^{2-} \quad \frac{\begin{array}{l}8 \text{ protons} = \quad 8+ \\ 10 \text{ electrons} = 10-\end{array}}{\text{Overall charge} = \quad 2-}$$

The process of gaining or losing electrons is known as **ionisation**.

Table 2.10 shows some common ions. You will notice that:

» some ions contain more than one type of atom, for example NO_3^-
» an ion may possess more than one unit of charge (either negative or positive), for example, Al^{3+}, O^{2-} and SO_4^{2-}.

❓ Worked example

1 Show how the sodium ions shown in Table 2.10 are formed.

$$_{11}Na^+ \quad \frac{\begin{array}{l}11 \text{ protons} = 11+ \\ 10 \text{ electrons} = 10-\end{array}}{\text{Overall charge} = \quad 1+}$$

2 Show how the sulfide ions shown in Table 2.10 are formed.

$$_{16}S^{2-} \quad \frac{\begin{array}{l}16 \text{ protons} = 16+ \\ 18 \text{ electrons} = 18-\end{array}}{\text{Overall charge} = \quad 2-}$$

▼ **Table 2.10** Some common ions

Name	Formula
Lithium ion	Li^+
Sodium ion	Na^+
Potassium ion	K^+
Magnesium ion	Mg^{2+}
Calcium ion	Ca^{2+}
Aluminium ion	Al^{3+}
Zinc ion	Zn^{2+}
Ammonium ion	NH_4^+
Fluoride ion	F^-
Chloride ion	Cl^-
Bromide ion	Br^-
Hydroxide ion	OH^-
Oxide ion	O^{2-}
Sulfide ion	S^{2-}
Carbonate ion	CO_3^{2-}
Nitrate ion	NO_3^-
Sulfate ion	SO_4^{2-}

▶ Test yourself

14 a State three differences between an electron and a proton.
   b State two similarities between a neutron and a proton.
   c Why are atoms electrically neutral?
15 Copy and complete the following table by writing the symbol for the element shown from the given number of particles present. The first one is done for you.

Element	Symbol	Particles present
Nitrogen	$^{14}_{7}N$	7p, 7n, 7e
Aluminium		13p, 14n, 13e
Potassium		19p, 20n, 19e
Argon		18p, 22n, 18e

16 Copy and complete the following table giving the charge on the ions shown.

Element	Proton number	Number of electrons	Charge on ion
Sodium	11	10	
Fluorine	9	10	
Magnesium	12	10	

Isotopes

In some elements, not all of the atoms in a sample of the element are identical. Some atoms of the same element can contain different numbers of neutrons and so have different mass numbers. Atoms of the same element which have different neutron numbers are called **isotopes**. The two isotopes of chlorine are shown in Figure 2.15.

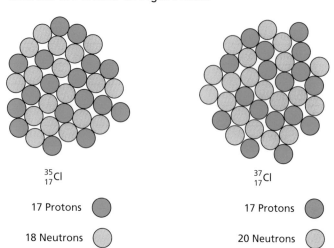

$^{35}_{17}Cl$

17 Protons

18 Neutrons

$^{37}_{17}Cl$

17 Protons

20 Neutrons

▲ **Figure 2.15** The nuclei of two isotopes of chlorine

Key definition

Isotopes are atoms of the same element that have the same number of protons but different numbers of neutrons.

Isotopes of the same element have the same chemical properties because they have the same number of electrons and therefore the same electronic configuration (p. 26).

The only effect of the extra neutron is to alter the mass of the atom and any properties that depend on it, such as density. Other examples of atoms and their isotopes are shown in Table 2.11.

There are two types of isotopes: those which are stable and those which are unstable. The isotopes which are unstable, as a result of the extra neutrons in their nuclei, are **radioactive** and are called **radioisotopes**. The nuclei of these atoms break up spontaneously with the release of not only large amounts of energy, but also certain types of dangerous radiations that can in some cases be useful to society. For example, uranium-235 is used as a source of power in nuclear reactors in nuclear power stations and cobalt-60 is used in radiotherapy treatment in hospitals (Figure 2.16); these are both radioisotopes.

▼ **Table 2.11** Some atoms and their isotopes

Element	Symbol	Particles present
Hydrogen	$^{1}_{1}H$	1e, 1p, 0n
(Deuterium)	$^{2}_{1}H$	1e, 1p, 1n
(Tritium)	$^{3}_{1}H$	1e, 1p, 2n
Carbon	$^{12}_{6}C$	6e, 6p, 6n
	$^{13}_{6}C$	6e, 6p, 7n
	$^{14}_{6}C$	6e, 6p, 8n
Oxygen	$^{16}_{8}O$	8e, 8p, 8n
	$^{17}_{8}O$	8e, 8p, 9n
	$^{18}_{8}O$	8e, 8p, 10n
Strontium	$^{86}_{38}Sr$	38e, 38p, 48n
	$^{88}_{38}Sr$	38e, 38p, 50n
	$^{90}_{38}Sr$	38e, 38p, 52n
Uranium	$^{235}_{92}U$	92e, 92p, 143n
	$^{238}_{92}U$	92e, 92p, 146n

▲ **Figure 2.16** Cobalt-60 is used in radiotherapy treatment. A beam of gamma rays produced by the radioactive isotope is directed into the patient's body to kill tumour tissue.

Going further

The mass spectrometer

How do we know isotopes exist? They were first discovered by scientists using apparatus called a mass spectrometer (Figure 2.17). The first mass spectrometer was built by the British scientist Francis Aston in 1919 and enabled scientists to compare the relative masses of atoms accurately for the first time.

▲ **Figure 2.17** A diagram of a mass spectrometer

A vacuum exists inside the mass spectrometer. A sample of the vapour of the element is injected into the ionisation chamber where it is bombarded by electrons. The collisions which take place between these electrons and the injected atoms cause an electron to be lost from the atom, which becomes a positive ion with a +1 charge. These positive ions are then accelerated towards a negatively charged plate, in the acceleration area.

The spectrometer is set up to ensure that when the ions leave the acceleration area they all have the same kinetic energy, regardless of the mass of the ions. This means that the lighter ions travel faster than the heavier ones, and effectively separate the ions according to their mass. Having left the acceleration area, the time for the ions to reach the detector is recorded. The detector counts the number of each of the ions which fall upon it and so a measure of the percentage abundance of each isotope is obtained. A typical mass spectrum for chlorine is shown in Figure 2.18.

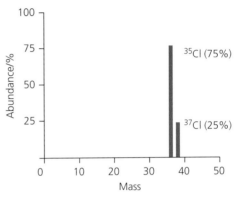

▲ **Figure 2.18** The mass spectrum for chlorine

Relative atomic mass

The average mass of a large number of atoms of an element is called its **relative atomic mass (A_r)**. This quantity takes into account the percentage abundance of all the isotopes of an element which exist.

> **Key definition**
>
> **Relative atomic mass**, A_r, is the average mass of the isotopes of an element compared to 1/12th of the mass of an atom of ^{12}C.

In 1961 the International Union of Pure and Applied Chemistry (IUPAC) recommended that the standard used for the A_r scale was carbon-12. An atom of carbon-12 was taken to have a mass of 12 amu. The A_r of an element is the average mass of the naturally occurring atoms of an element on a scale where ^{12}C has a mass of exactly 12 units.

$$A_r = \frac{\text{average mass of isotopes of the element}}{\frac{1}{12} \times \text{mass of one atom of carbon-12}}$$

Note: $\frac{1}{12}$ of the mass of one carbon-12 atom = 1 amu.

❓ Worked example

What is the relative atomic mass of chlorine? Chlorine has two isotopes:

	$^{35}_{17}Cl$	$^{37}_{17}Cl$
% abundance	75	25

Hence the 'average mass' or A_r of a chlorine atom is:

$$\frac{(75 \times 35) + (25 \times 37)}{100} = 35.5$$

$$A_r = \frac{35.5}{1}$$
$$= 35.5 \text{ amu}$$

▶ Test yourself

17 Calculate the number of neutrons in the following atoms:

a $^{27}_{13}Al$ b $^{31}_{15}P$ c $^{91}_{40}Zr$ d $^{190}_{76}Os$.

18 Given that the percentage abundance of $^{20}_{10}Ne$ is 90% and that of $^{22}_{10}Ne$ is 10%, calculate the A_r of neon.

The arrangement of electrons in atoms

The nucleus of an atom contains the heavier sub-atomic particles – the protons and the neutrons. The electrons, the lightest of the sub-atomic particles, move around the nucleus at great distances from the nucleus relative to their size. They move very fast in electron shells, very much like the planets orbit the Sun.

It is not possible to give the exact position of an electron in an electron shell. However, we can state that electrons can only occupy certain, definite electron shells and that they cannot exist between them. Also, as mentioned earlier, each of the electron shells can hold only a certain number of electrons.

» First shell holds up to two electrons.
» Second shell holds up to eight electrons.
» Third shell holds up to 18 electrons.

There are further electron shells which contain increasing numbers of electrons.

The third electron shell can be occupied by a maximum of 18 electrons. However, when eight electrons have occupied this shell, a certain stability is given to the atom and the next two electrons go into the fourth electron shell, and then the remaining ten electrons complete the third shell.

The electrons fill the electron shell starting from the shell nearest to the nucleus, which has the lowest energy. When this is full (with two electrons), the next electron goes into the second electron shell. When this shell is full with eight electrons, then the electrons begin to fill the third and fourth shells as stated above.

For example, a $^{16}_{8}O$ atom has a proton number of 8 and therefore has eight electrons. Two of the eight electrons enter the first electron shell, leaving six to occupy the second electron shell, as shown in Figure 2.19. The electronic configuration for oxygen can be written in a shorthand way as 2,6.

There are 118 elements, and Table 2.12 shows the way in which the electrons are arranged in the first 20 of these elements. The way in which the electrons are distributed is called the **electronic configuration** (electronic structure). Figure 2.20 shows the electronic configuration of a selection of atoms.

▲ **Figure 2.19** Arrangement of electrons in an oxygen atom

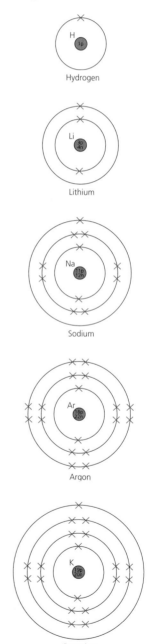

▲ **Figure 2.20** Electronic configurations of hydrogen, lithium, sodium, argon and potassium

▼ **Table 2.12** Electronic configuration of the first 20 elements

Element	Symbol	Proton number	Number of electrons	Electronic configuration
Hydrogen	H	1	1	1
Helium	He	2	2	2
Lithium	Li	3	3	2,1
Beryllium	Be	4	4	2,2
Boron	B	5	5	2,3
Carbon	C	6	6	2,4
Nitrogen	N	7	7	2,5
Oxygen	O	8	8	2,6
Fluorine	F	9	9	2,7
Neon	Ne	10	10	2,8
Sodium	Na	11	11	2,8,1
Magnesium	Mg	12	12	2,8,2
Aluminium	Al	13	13	2,8,3
Silicon	Si	14	14	2,8,4
Phosphorus	P	15	15	2,8,5
Sulfur	S	16	16	2,8,6
Chlorine	Cl	17	17	2,8,7
Argon	Ar	18	18	2,8,8
Potassium	K	19	19	2,8,8,1
Calcium	Ca	20	20	2,8,8,2

From Table 2.12 you can see helium has a full outer shell of two electrons, and neon has a full outer shell of 8 electrons. These two elements are very unreactive for this reason. In Chapter 9 we will see that the number of outer shell electrons is related to the position in the Periodic Table. Helium and neon are part of a group of elements known as the noble or inert gases and they are generally very stable and unreactive (p. 145). This is linked to their full outer shells. When elements react to form compounds, they do so to achieve full electron shells, and this idea forms the basis of the electronic theory of chemical bonding, which we will discuss further in the next chapter.

> ## ▶ Test yourself
>
> 19 How many electrons may be accommodated in each of the first three electron shells?
> 20 What is the same about the electronic configurations of:
>   a lithium, sodium and potassium?
>   b beryllium, magnesium and calcium?
> 21 An element X has a proton number of 13. What is the electronic configuration of X?

Revision checklist

After studying Chapter 2 you should be able to:
- ✔ Describe the differences between elements, compounds and mixtures.
- ✔ Interpret and use symbols for given atoms.
- ✔ State the formulae of the elements and compounds you have dealt with.
- ✔ Define the molecular formula of a compound as the number and type of different atoms in one molecule.
- ✔ Deduce the formula of a simple compound from the relative numbers of atoms present in a model or a diagrammatic representation of the compound.
- ✔ Construct word equation and symbol equations to show how reactants form products, including state symbols.
- ✔ Define redox reactions as involving both oxidation and reduction.
- ✔ Define oxidation as oxygen gain and reduction as oxygen loss.
- ✔ Identify redox reactions as reactions involving gain and loss of oxygen.

- ✔ Define an oxidising agent and a reducing agent.

- ✔ Describe the structure of the atom as a central nucleus containing neutrons and protons surrounded by electrons in shells.
- ✔ State the relative charges and relative masses of a proton, a neutron.
- ✔ Define proton number and atomic number as well as mass number.
- ✔ Determine the electronic configuration of elements with the proton number 1 to 20.
- ✔ Describe the formation of positive ions, known as cations, and negative ions, known as anions.
- ✔ State what isotopes are.

- ✔ State that isotopes of the same element have the same electronic configuration and so have the same chemical properties.
- ✔ Calculate the relative atomic mass (A_r) of an element from given data of the relative masses and abundance of their different isotopes.

Exam-style questions

1 a Define the terms:
  i proton [3]
  ii neutron [2]
  iii electron. [3]
 b An atom **X** has a proton number of 19 and relative atomic mass of 39.
  i How many electrons, protons and neutrons are there in atom **X**? [3]
  ii How many electrons will there be in the outer electron shell of atom **X**? [1]
  iii What is the electronic configuration of atom **X**? [1]

2 a $^{69}_{31}$Ga and $^{71}_{31}$Ga are isotopes of gallium. Use this example to explain what you understand by the term isotope. [3]
 b A sample of gallium is 60% $^{69}_{31}$Ga atoms and 40% $^{71}_{31}$Ga atoms. Calculate the relative atomic mass of this sample of gallium. [2]

3 Define the following terms using specific examples to help with your explanation:
 a element [2]
 b metal [3]
 c non-metal [2]
 d compound [2]
 e molecule [2]
 f mixture. [2]

4 State which of the substances listed below are:
 a metallic elements
 b non-metallic elements
 c compounds
 d mixtures.
   silicon, sea water, calcium, argon, water, air, carbon monoxide, iron, sodium chloride, diamond, brass, copper, dilute sulfuric acid, sulfur, oil, nitrogen, ammonia [17]

5 State, at room temperature and pressure (r.t.p.), which of the substances listed below is/are:
 a a solid element
 b a liquid element
 c a gaseous mixture
 d a solid mixture
 e a liquid compound
 f a solid compound.
   bromine, carbon dioxide, helium, steel, air, oil, marble, copper, water, sand, tin, bronze, mercury, salt [11]

6 a How many atoms of the different elements are there in the formulae of these compounds?
  i nitric acid, HNO_3 [3]
  ii methane, CH_4 [2]
  iii copper nitrate, $Cu(NO_3)_2$ [3]
  iv ethanoic acid, CH_3COOH [3]
  v sugar, $C_{12}H_{22}O_{11}$ [3]
  vi phenol, C_6H_5OH [3]
  vii ammonium sulfate, $(NH_4)_2SO_4$ [4]
 b Balance the following equations.
  i $Zn(s) + O_2(g) \rightarrow ZnO(s)$ [2]
  ii $Fe(s) + Cl_2(g) \rightarrow FeCl_3(s)$ [3]
  iii $Li(s) + O_2(g) \rightarrow Li_2O(s)$ [2]
  iv $H_2(g) + O_2(g) \rightarrow H_2O(g)$ [2]
  v $Mg(s) + CO_2(g) \rightarrow MgO(s) + C(s)$ [2]

Bonding and structure

FOCUS POINTS

★ In what different ways do elements combine to form compounds?
★ How do the bonds within a compound affect its properties?
★ What is a redox reaction?

In the previous chapter, you established that both metal and non-metal elements are made up of atoms, which can combine to form compounds. In this chapter you will look more closely at how different elements combine, or bond, to form compounds. You will see that, generally, when elements combine they can form either ionic bonds with one another and are known as ionic compounds, or they can form covalent bonds with one another and are known as covalent compounds. You will also look at a third type of bonding, found in metals only, called metallic bonding.

You will find out how the type of bonding affects the properties of these compounds. You will also learn more about formulae and how they can be obtained from given data and you will obtain an understanding of redox reactions and the processes associated with them.

▲ **Figure 3.1** The properties of salt are very different from those of the sodium and chlorine it was made from. To get salt into your diet you would not eat sodium or inhale chlorine!

3.1 Ionic bonding

Ionic bonds are usually found in compounds that contain metals combined with non-metals. When this type of bond is formed, electrons are transferred from the metal atoms to the non-metal atoms during the chemical reaction. In doing this, the atoms become more stable due to their full outer shells. For example, consider what happens when sodium and chlorine react together and combine to make sodium chloride (Figure 3.1).

sodium + chlorine → sodium chloride

Sodium has just one electron in its outer shell ($_{11}$Na 2,8,1). Chlorine has seven electrons in its outer shell ($_{17}$Cl 2,8,7). When these two elements react, the outer electron of each sodium atom is transferred to the outer shell of a chlorine atom (Figure 3.2).

In this way both the atoms obtain full outer shells and become 'like' the noble gas nearest to them in the Periodic Table (see Figure 9.3, p. 135). One way of showing this transfer of electrons is by using a dot-and-cross diagram. In these diagrams, the electrons from one atom are shown as crosses and those of the other atom as dots. In the reaction the sodium atom has become a sodium ion, in a process known as ionisation. This sodium ion has an electronic configuration like the noble gas neon.

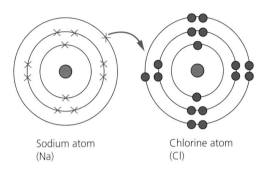

Sodium atom
(Na)

Chlorine atom
(Cl)

Sodium ion
(Na⁺)

Chloride ion
(Cl⁻)

11 protons = 11+
10 electrons = 10−
Overall charge = 1+

17 protons = 17+
18 electrons = 18−
Overall charge = 1−

▲ **Figure 3.2** Ionic bonding in sodium chloride

To lose electrons in this way is called **oxidation**. The chlorine atom has become a chloride ion with an electronic configuration like argon. To gain electrons in this way is called **reduction**.

In the chemical process producing sodium chloride, both oxidation and reduction have taken place and so this is known as a **redox reaction**.

Key definitions

Oxidation involves loss of electrons.

Reduction involves gain of electrons.

Redox reactions involve simultaneous oxidation and reduction.

A further discussion of oxidation and reduction in terms of electron transfer takes place in Chapter 5 (p. 71).

Only the outer electrons are important in bonding, so we can simplify the diagrams by missing out the inner shells (Figure 3.3).

The charges on the sodium and chloride ions are equal but opposite. They balance each other and the resulting formula for sodium chloride is NaCl. These oppositely charged ions attract each other and are pulled, or bonded, to one another by strong electrostatic forces. This type of bonding is called **ionic bonding**.

Key definition

An **ionic bond** is a strong electrostatic attraction between oppositely charged ions.

All students should be able to describe the formation of ionic bonds between elements of Group I and Group VII and draw dot-and-cross diagrams. Extended learners should be able to do this for any metallic and non-metallic element, including drawing dot-and-cross diagrams.

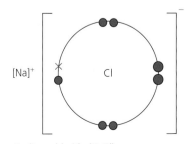

Sodium chloride (NaCl)

▲ **Figure 3.3** Simplified diagram of ionic bonding in sodium chloride

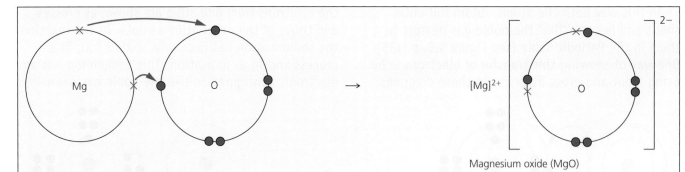

▲ **Figure 3.4** Simplified diagram of ionic bonding in magnesium oxide

Figure 3.4 shows the electron transfers that take place between a magnesium atom and an oxygen atom during the formation of magnesium oxide.

Magnesium obtains a full outer shell by losing two electrons. These are transferred to the oxygen atom. In magnesium oxide, the Mg^{2+} and O^{2-} are oppositely charged and are attracted to one another. The formula for magnesium oxide is MgO.

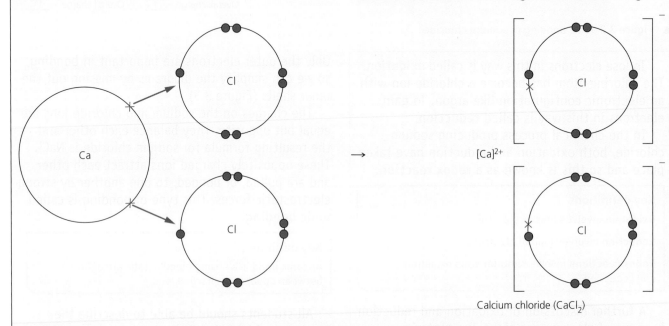

▲ **Figure 3.5** The transfer of electrons that takes place during the formation of calcium chloride

When calcium and chlorine react, the calcium atom gives each of the two chlorine atoms one electron (Figure 3.5). In this case, a compound is formed containing two chloride ions (Cl^-) for each calcium ion (Ca^{2+}). The chemical formula is $CaCl_2$.

Test yourself

1 Define the term 'ionic bond'.
2 Why do inert (noble) gases *not* bond with themselves and other elements?
3 Draw diagrams to represent the bonding in each of the following ionic compounds:
   a potassium fluoride (KF)
   b lithium chloride (LiCl)
   c magnesium fluoride (MgF_2)
   d calcium oxide (CaO).

➡ Going further

Scientists, using **X-ray diffraction** (Figure 3.6a), have obtained photographs that indicate the way in which the ions are arranged (Figure 3.6b). The electron density map of sodium chloride is shown in Figure 3.6c.

Figure 3.6d shows the structure of sodium chloride as determined by the X-ray diffraction technique. The study of crystals using X-ray diffraction was pioneered by Sir William Bragg and his son Sir Lawrence Bragg in 1912. X-rays are a form of electromagnetic radiation. They have a much shorter wavelength than light therefore it

is possible to use them to investigate extremely small structures.

When X-rays are passed through a crystal of sodium chloride, for example, you get a pattern of spots called a diffraction pattern (Figure 3.6b). This pattern can be recorded on photographic film and used to work out how the ions or atoms are arranged in the crystal. Crystals give particular diffraction patterns depending on their structure, and this makes X-ray diffraction a particularly powerful technique in the investigation of crystal structures.

a X-ray diffraction technique

c Electron density map of sodium chloride

b X-ray diffraction photograph of sodium chloride

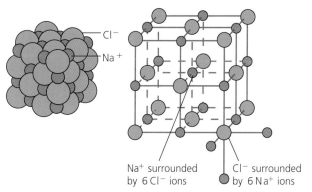

Na⁺ surrounded by 6 Cl⁻ ions   Cl⁻ surrounded by 6 Na⁺ ions

d The structure of sodium chloride

▲ **Figure 3.6**

Ionic structures

Ionic structures are solids at room temperature and have high melting and boiling points.

> The ions are packed together in a regular arrangement called a **lattice**. Within the lattice, oppositely charged ions attract one another strongly.
>
> Figure 3.6d shows only a tiny part of a small crystal of sodium chloride. Many millions of sodium ions and chloride ions would be arranged in this way in a crystal of sodium chloride to make up the **giant ionic lattice** structure. Each sodium ion in the lattice is surrounded by six chloride ions, and each chloride ion is surrounded by six sodium ions.

➡ Going further

Not all ionic substances form the same structures. Caesium chloride (CsCl), for example, forms a different structure due to the larger size of the caesium ion compared with that of the sodium ion. This gives rise to the structure shown in Figure 3.7, which is called a body-centred cubic structure. Each caesium ion is surrounded by eight chloride ions and, in turn, each chloride ion is surrounded by eight caesium ions.

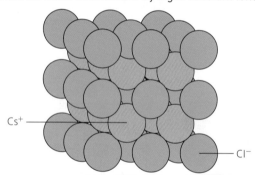

Cs⁺

Cl⁻

a The arrangement of ions

Cs⁺ surrounded by 8 Cl⁻ ions
In the same way, each Cl⁻
is surrounded by 8 Cs⁺ ions

b The unit cell

▲ **Figure 3.7** The structure of caesium chloride

Properties of ionic compounds

Ionic compounds have the following properties.

» They are usually solids at room temperature, with high melting and boiling points.

> This is due to the strong electrostatic forces holding the crystal lattice together. A lot of energy is therefore needed to separate the ions and melt the substance.

» They are usually hard substances.
» They mainly dissolve in water.

> This is because water molecules are able to bond with both the positive and the negative ions, which breaks up the lattice and keeps the ions apart. Figure 3.8 shows the interaction between water molecules (the solvent) and sodium and chloride ions from sodium chloride (the solute). For a further discussion of the solubility of substances see, Chapter 8, p. 120.

» They usually conduct electricity when in the molten state or in aqueous solution.

> The forces of attraction between the ions are weakened and the ions are free to move to the appropriate electrode. This allows an electric current to be passed through the molten compound (see Chapter 5, p. 71).

» They usually cannot conduct electricity when solid,

> because the ions are not free to move.

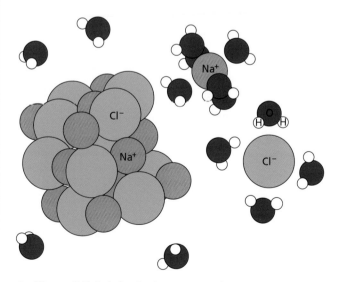

Na⁺

Cl⁻

O
H   H

Na⁺

Cl⁻

▲ **Figure 3.8** Salt (an ionic compound) dissolving in water

Test yourself

4 Link the terms in the boxes on the left with the definitions on the right.

Oxidation	An atom or group of atoms which has lost one or more electrons
Reduction	Involves loss of electrons
Redox	A strong electrostatic force of attraction between oppositely charged ions
Negative ion	Involves gain of electrons
Positive ion	Involves both oxidation and reduction
Ionic bond	An atom or group of atoms which has gained one or more electrons

Formulae of ionic substances

We saw previously (see p. 31) that ionic compounds contain positive and negative ions, whose charges balance. For example, sodium chloride contains one Na^+ ion for every Cl^- ion, giving rise to the formula NaCl. This method can be used to write down formulae which show the ratio of the number of ions present in any ionic compound.

The formula of magnesium chloride is $MgCl_2$. This formula is arrived at by each Mg^{2+} ion combining with two Cl^- ions, and once again the charges balance. The use of oxidation numbers is useful when working out the formula of a compound. An oxidation number shows how oxidised or reduced an ion is compared to its atom. Na^+ has an oxidation number of +1 because it is formed by the loss of one electron (Chapter 5, p. 71) from a sodium atom but Mg^{2+} has an oxidation number of +2 because it is formed when a magnesium atom loses two electrons (Table 3.1). Na^+ can bond (combine) with only one Cl^- ion, whereas Mg^{2+} can bond with two Cl^- ions. A chloride ion has an oxidation state of –1 because it is formed when a chlorine atom gains one extra electron.

Some elements, such as copper and iron, possess two ions with different oxidation numbers. Copper can form the Cu^+ ion and the Cu^{2+} ion, with oxidation numbers of 1 and 2 respectively. Therefore it can form two different compounds with chlorine, CuCl and $CuCl_2$. We can also distinguish the difference by using Roman numerals in their names: CuCl is copper(I) chloride and $CuCl_2$ is copper(II) chloride. Similarly, iron forms the Fe^{2+} and Fe^{3+} ions and so can also form two different compounds with, for example, chlorine: $FeCl_2$ (iron(II) chloride) and $FeCl_3$ (iron(III) chloride).

Table 3.1 shows the oxidation numbers of a series of ions you will normally meet in your study of chemistry.

You will notice that Table 3.1 includes groups of atoms which have net charges. For example, the nitrate ion is a single unit composed of one nitrogen atom and three oxygen atoms, and has one single negative charge. The formula, therefore, of magnesium nitrate would be $Mg(NO_3)_2$. You will notice that the NO_3 has been placed in brackets with a $_2$ outside the bracket. This indicates that there are two nitrate ions present for every magnesium ion. The ratio of the atoms present is therefore:

Mg (N O$_3$)$_2$
1Mg : 2N : 6O

▼ Table 3.1 Oxidation numbers (valencies) of some elements (ions) and groups of atoms

| | Oxidation numbers (valencies) | | | | | |
	1		2		3	
Metals	Lithium	(Li^+)	Magnesium	(Mg^{2+})	Aluminium	(Al^{3+})
	Sodium	(Na^+)	Calcium	(Ca^{2+})	Iron(III)	(Fe^{3+})
	Potassium	(K^+)	Copper(II)	(Cu^{2+})		
	Silver	(Ag^+)	Zinc	(Zn^{2+})		
	Copper(I)	(Cu^+)	Iron(II)	(Fe^{2+})		
			Lead	(Pb^{2+})		
			Barium	(Ba^{2+})		

	Oxidation numbers (valencies)					
	1		**2**		**3**	
Non-metals	Fluoride	(F^-)	Oxide	(O^{2-})		
	Chloride	(Cl^-)	Sulfide	(S^{2-})		
	Bromide	(Br^-)				
	Hydrogen	(H^+)				
Groups of atoms	Hydroxide	(OH^-)	Carbonate	(CO_3^{2-})	Phosphate	(PO_4^{3-})
	Nitrate	(NO_3^-)	Sulfate	(SO_4^{2-})		
	Ammonium	(NH_4^+)	Dichromate(VI)	$(Cr_2O_7^{2-})$		
	Hydrogencarbonate	(HCO_3^-)				
	Manganate (VII)	(MnO_4^-)				

Oxidation number

Each atom in an element or compound is assigned an oxidation number to show how much it is reduced or oxidised. The following points should be remembered when using oxidation numbers.

» Roman numerals (I, II, III, IV, V, VI, VII, VIII) are used in writing the oxidation number of an element.

» This number is placed after the element that it refers to. For example, the name for $FeCl_3$ is iron(III) chloride and not iron(3) chloride.
» The oxidation number of the free element is always 0, for example, in metals such as zinc and copper.
» In simple monatomic ions, the oxidation number is the same as the charge on the ion. So iodine has an oxidation number of 0 in I_2 but an oxidation number of −1 in I^-.
» Compounds have no charge overall. Hence the oxidation numbers of all the individual elements in a compound must add up to 0. The oxidation numbers of elements in compounds can vary, as seen in Table 3.1. It is possible to recognise which of the different oxidation numbers a metal element is in by the colour of its compounds (Figure 3.9).

▲ **Figure 3.9** Iron(II) sulfate is pale green, while iron(III) sulfate is yellow

» An increase in the oxidation number, for example, from +2 to +3 as in the case of Fe^{2+} to Fe^{3+}, is **oxidation**.
» However, a reduction in the oxidation number, for example, from +6 to +3 as in the case of Cr^{6+} (in CrO_4^{2-}) to Cr^{3+}, is **reduction**.

Key definitions

Oxidation involves an increase in oxidation number.

Reduction involves a decrease in oxidation number.

During a redox reaction, the substance that brings about oxidation is called an oxidising agent and is itself reduced during the process. A substance that brings about reduction is a reducing agent and is itself oxidised during the process.

For example, if a dilute solution of acidified potassium manganate(VII) (pale purple) is added to a solution of iron(II) sulfate, a colour change takes place as the reaction occurs (Figure 3.10). The iron(II) sulfate (pale green) changes to pale yellow, showing the presence of iron(III) ions.

▲ **Figure 3.10** Manganate(VII) ions (oxidising agent) and iron(II) ions (reducing agent) are involved in a redox reaction when mixed

In this reaction, the iron(II) ions have been oxidised to iron(III) ions (increase in oxidation number) and the manganate(VII) ions have been reduced to manganese(II) ions (decrease in oxidation number), which are very pale pink. Hence the manganate(VII) ions are the oxidising agent and the iron(II) ions are the reducing agent.

Potassium iodide is a common reducing agent. When it is oxidised, a colour change is also produced. The iodide ion (I^-) is oxidised to iodine(I_2). The colour of the resulting solution will change from colourless to yellow-brown. If you then add starch indicator, a test to show the presence of iodine, it will turn blue-black. Hence if potassium iodide solution is added to a solution of iron(III) solution followed by the starch indicator, then a blue-black colour will be seen as shown in Figure 3.11. This shows that the iron(III) ions must have been reduced to iron(II) ions as the iodide ions are oxidised to iodine.

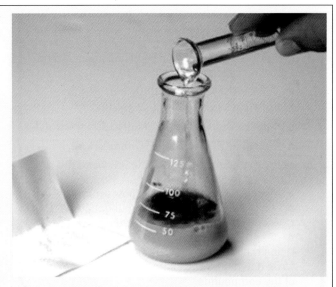

▲ **Figure 3.11** The blue-black colour shows the presence of iron(III) ions

It is possible to recognise redox processes by looking at the oxidation numbers on the two sides of the chemical equation for a reaction. For example, in the reaction between magnesium and dilute sulfuric acid, the magnesium dissolves and hydrogen gas is produced. Both magnesium metal and hydrogen gas are free elements and so have an oxidation number of 0. In sulfuric acid, hydrogen has an oxidation number of +1 since the overall charge on the sulfate ion is −2. Similarly, the oxidation number of magnesium in magnesium sulfate is +2.

$$\text{magnesium} + \text{sulfuric} \rightarrow \text{magnesium} + \text{hydrogen}$$
$$\text{acid} \qquad\qquad \text{sulfate}$$

$$\underset{0}{Mg(s)} + \underset{+1}{H_2SO_4(aq)} \rightarrow \underset{+2}{MgSO_4(aq)} + \underset{0}{H_2(g)}$$

The sulfate ions are unchanged by the reaction and so can be ignored.

As you can see, the oxidation number of magnesium has increased from 0 to +2. Therefore the magnesium has been oxidised by the sulfuric acid and so sulfuric acid is the **oxidising agent**. The oxidation number of hydrogen has decreased from +1 in the sulfuric acid to 0 in the free element. Therefore the hydrogen has been reduced by the magnesium and so magnesium is the **reducing agent**.

Key definitions

An **oxidising agent** is a substance that oxidises another substance and is itself reduced.

A **reducing agent** is a substance that reduces another substance and is itself oxidised.

Going further

The 'cross-over' method

A less scientific but simpler method to work out the formulae of compounds is called the 'cross-over' method. In this method it is only necessary to 'swap' the numerical value of the oxidation number of the elements or groups of atoms (or radicals) concerned. This is easily done by 'crossing over' the numerical value of their oxidation numbers, by placing the numerical value of the oxidation number of the first element after the symbol of the second and placing the numerical value of the oxidation number of the second element or radical after the symbol of the first.

For example, in aluminium sulfide the oxidation numbers of the elements are:

$$Al = 3 \quad \text{and} \quad S = 2$$

Hence the chemical formula is Al_2S_3.

Similarly, in sodium sulfate the oxidation numbers are:

$$Na = 1 \quad \text{and} \quad SO_4 = 2$$

Hence the chemical formula is Na_2SO_4.

3.2 Covalent bonding

Another way in which atoms can form stable compounds is by sharing the electrons in their outer shells. This occurs between non-metal atoms, and the bond formed is called a **covalent bond**. During the bond formation, the atoms involved gain the stability of the noble (inert) gas electronic configuration. The simplest example of this type of bonding can be seen by considering the hydrogen molecule, H_2.

> **Key definition**
> A **covalent bond** is formed when a pair of electrons is shared between two atoms leading to noble gas electronic configurations.

Each hydrogen atom in the molecule has one electron. In order to obtain a full outer shell and gain an electronic configuration that is the same as the noble gas helium, each of the hydrogen atoms must have two electrons. To do this, the outer shells of the two hydrogen atoms overlap (Figure 3.12a). A molecule of hydrogen is formed, with two hydrogen atoms sharing a pair of electrons (Figure 3.12a, b). This shared pair of electrons is known as a single covalent bond and is represented by a single line as in hydrogen:

H—H

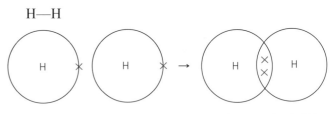

2 hydrogen atoms   Hydrogen molecule (H_2)

a The electron sharing to form the single covalent bond in H_2 molecules

b Model of a hydrogen molecule

▲ **Figure 3.12**

A similar example exists in the diatomic halogen molecule chlorine, Cl_2 (Figure 3.13).

2 chlorine atoms

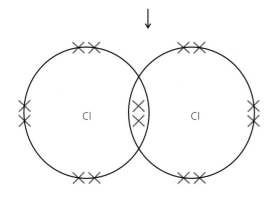

Chlorine molecule (Cl$_2$)

▲ **Figure 3.13** The electron sharing to form the single covalent bond in Cl$_2$ molecules (sharing outer electron shells only)

Another simple molecule is hydrogen chloride. This time two different elements share electrons to gain the electronic configuration of their nearest noble gas. In this case hydrogen, with only one electron in its outer shell, needs to share one more electron to gain the electronic configuration of helium. Chlorine has seven electrons in its outer shell so it also needs to share one more electron, this time to gain the electronic configuration of the noble gas argon (Figure 3.14).

▲ **Figure 3.14** Dot-and-cross diagram to show the formation of a hydrogen chloride molecule.

Other covalent compounds

Methane (natural gas) is a gas whose molecules contain atoms of carbon and hydrogen. The electronic configurations are:

$$_6C \; 2,4 \qquad _1H \; 1$$

The carbon atom needs four more electrons to attain the electronic configuration of the noble gas neon. Each hydrogen atom needs only one electron to form the electronic configuration of helium. Figure 3.15 shows how the atoms gain these electronic configurations by the sharing of electrons. Note that only the outer electron shells are shown.

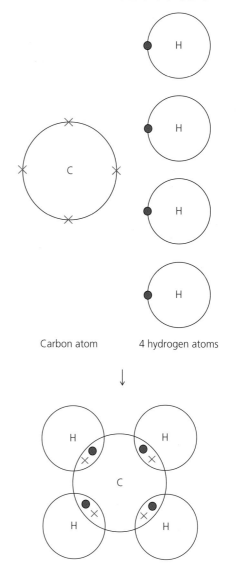

Carbon atom   4 hydrogen atoms

Methane molecule (CH$_4$)

▲ **Figure 3.15** Dot-and-cross diagram to show the formation of a methane molecule

▲ **Figure 3.16** Model of a methane molecule

Ammonia is a gas containing the elements nitrogen and hydrogen. It is used in large amounts to make fertilisers. The electronic configurations of the two elements are:

$_7$N 2,5 $_1$H 1

The nitrogen atom needs three more electrons to obtain the noble gas structure of neon. Each hydrogen requires only one electron to form the noble gas structure of helium. The nitrogen and hydrogen atoms share electrons, forming three single covalent bonds (Figure 3.17).

Water is a liquid containing the elements hydrogen and oxygen. The electronic configurations of the two elements are:

$_8$O 2,6 $_1$H 1

The oxygen atom needs two electrons to gain the electronic configuration of neon. Each hydrogen requires one more electron to gain the electronic configuration of helium. Again, the oxygen and hydrogen atoms share electrons, forming a water molecule with two single covalent bonds as shown in Figure 3.19.

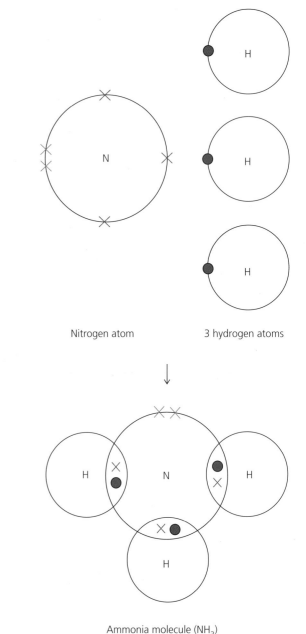

Nitrogen atom   3 hydrogen atoms

Ammonia molecule (NH_3)

▲ **Figure 3.17** Dot-and-cross diagram to show the formation of an ammonia molecule

▲ **Figure 3.18** Model of the ammonia molecule

▲ **Figure 3.20** Model of a water molecule

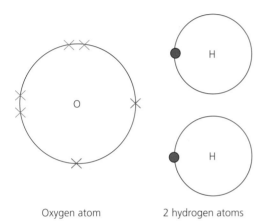

Oxygen atom   2 hydrogen atoms

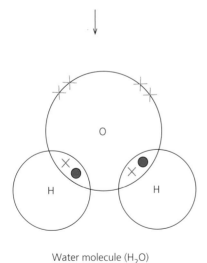

Water molecule (H_2O)

▲ **Figure 3.19** Dot-and-cross diagram to show the formation of a water molecule

Carbon dioxide is a gas containing the elements carbon and oxygen. The electronic configurations of the two elements are:

$_6C$ 2,4 $_8O$ 2,6

In this case each carbon atom needs to share four electrons to gain the electronic configuration of neon. Each oxygen needs to share two electrons to gain the electronic configuration of neon. This is achieved by forming two double covalent bonds in which two pairs of electrons are shared in each case, as shown in Figure 3.21. Carbon dioxide is a linear molecule (Figure 3.22).

2 oxygen atoms   Carbon atom

Carbon dioxide molecule (CO_2)

▲ **Figure 3.21** Dot-and-cross diagram to show the formation of a carbon dioxide molecule

41

O═C═O

a Carbon dioxide molecule. Note the double covalent bond is represented by a double line

b Model of a linear carbon dioxide molecule

▲ **Figure 3.22**

Another molecule which contains a double covalent bond is that of oxygen gas, O_2. Each of the oxygen atoms has six electrons in its outer electron shell.

$_8$O 2,6

Each needs to share another two electrons to gain the electronic configuration of the noble gas neon.

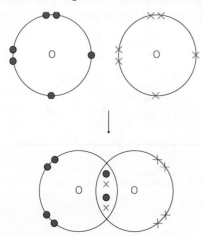

▲ **Figure 3.23** Dot-and-cross diagram to show the formation of an oxygen molecule

As with the carbon dioxide molecule, the double bond can be shown by a double line between the two oxygen atoms, representing two shared pairs of electrons.

O═O

Nitrogen gas, N_2, is molecule which contains a triple covalent bond. A triple covalent bond is formed when three pairs of electrons are shared and it is represented by '≡'.

Each nitrogen atom has the electronic configuration shown below:

$_7$N 2,5

Both nitrogen atoms in the molecule need to share a further three electrons to gain the electronic configuration of the noble gas neon.

The triple bond can be shown in the molecule as N≡N.

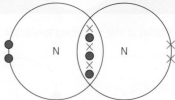

▲ **Figure 3.24** Dot-and-cross diagram to show the bonding in a nitrogen molecule

Methanol, CH_3OH, is a molecule which contains three different types of atom. When this happens and you need to draw a dot-and-cross diagram, simply make sure that atoms which bond with one another do not both have dots or crosses. In this molecule the atoms have these configurations:

$_6$C 2,4 $_1$H 1 $_8$O 2,6

The carbon atom needs to share four further electrons which it can do with the three hydrogens; each needs to share one more electron, and the oxygen atom. The oxygen atom, which shares one electron with the carbon atom, shares another electron with the remaining hydrogen atom. By doing this, the hydrogen atoms gain the electronic configuration of helium, and the carbon and oxygen atoms gain the electronic configuration of neon.

The dot-and-cross diagram for methanol is:

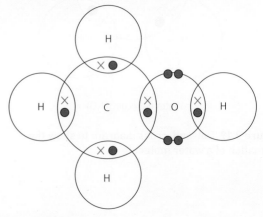

▲ **Figure 3.25** Dot-and-cross diagram to show the formation of a methanol molecule

Test yourself

8 What do you understand by the term 'covalent bond'?

9 Draw diagrams to represent the bonding in each of the following covalent compounds:
   a tetrachloromethane (CCl_4)
   b hydrogen sulfide (H_2S)

Covalent structures

Compounds containing covalent bonds have molecules whose structures can be classified as either **simple molecular** or **giant covalent**.

Simple molecular structures are simple, formed from only a few atoms. They have strong covalent bonds between the atoms within a molecule (**intramolecular bonds**) but have weak bonds between the molecules (**intermolecular forces**).

 Going further

Some of the strongest of these weak intermolecular forces occur between water molecules.

One type of weak bond between molecules is known as the van der Waals bond (or force), and these forces increase steadily with the increasing size of the molecule.

Examples of simple molecules are iodine (Figure 3.26), methane, water and ethanol.

Weak intermolecular bond

Strong covalent bond

▲ **Figure 3.26** Strong covalent and weak intermolecular forces in iodine

Giant covalent structures contain many hundreds of thousands of atoms joined by strong covalent bonds. Examples of substances with this type of structure are diamond, graphite and silicon(IV) oxide (Figure 3.27).

Properties of covalent compounds

Covalent compounds have the following properties.
» As simple molecular compounds, they are usually gases, liquids or solids with low melting and boiling points.

The melting points are low because of the weak intermolecular forces of attraction which exist between simple molecules. These are weaker compared to the strong covalent bonds. Giant covalent substances have higher melting points, because the whole structure is held together by strong covalent bonds.

It should be noted that in ionic compounds, the interionic forces are much stronger than the intermolecular forces in simple covalent substances and so the melting and boiling points are generally higher.
» Generally, they do not conduct electricity when molten or dissolved in water. This is because they do not contain ions. However, some molecules react with water to form ions. For example, hydrogen chloride gas produces aqueous hydrogen ions and chloride ions when it dissolves in water:

$$HCl(g) \xrightarrow{\text{water}} H^+(aq) + Cl^-(aq)$$

The presence of the ions allows the solution to conduct electricity.
» Generally, they do not dissolve in water. However, water is an excellent solvent and can interact with and dissolve some covalent molecules better than others. Covalent substances are generally soluble in organic solvents.

O

 Si

a The silicon(IV) oxide structure in quartz

▲ **Figure 3.27**

b Quartz is a hard solid at room temperature. It has a melting point of 1610°C and a boiling point of 2230°C

Going further

Giant covalent structures

When an element can exist in more than one physical form in the same physical state, it is said to exhibit allotropy (or polymorphism). Each of the different physical forms is called an allotrope. Allotropy is actually quite a common feature of the elements in the Periodic Table (p. 135). An example of an element which shows allotropy is carbon.

Practical skills

Testing the electrical conductivity of ionic and covalent substances

For safe experiments/demonstrations which are related to this chapter, please refer to the *Cambridge IGCSE Chemistry Practical Skills Workbook*, which is also part of this series.

Safety

- Eye protection must be worn.
- Take care when handling dilute aqueous solutions: most are low hazard (and should all be ≤1M); 0.2–1.0 mol/dm^3 copper(II) sulfate solution is an irritant and corrosive.

The apparatus was set up as shown in the diagram on the right to investigate the electrical conductivity of a number of compounds in aqueous solution.

1 Describe the experimental method you could use to carry out a fair test on a variety of ionic and covalent substances dissolved in water and using the apparatus shown in the diagram.

Solution	Does the bulb light?	Does the substance in solution conduct electricity?	Does the substance in solution contain ionic or covalent bonding?
Glucose ($C_6H_{12}O_6$)	No		
Potassium chloride	Yes		
Calcium chloride	Yes		
Sodium iodide	Yes		
Ethanol (C_2H_5OH)	No		
Copper(II) sulfate	Yes		
Pure water	No		

2 Copy and complete the results table above.
3 Using your answers to columns three and four in the results table, write a conclusion for this experiment. In your conclusion, state and explain any trends shown in the results.
4 Would the results be different if solid potassium chloride were used instead of potassium chloride solution? Explain your answer.

5 Predict and explain the results you would obtain for magnesium nitrate solution.
6 Predict and explain the results you would obtain with just distilled water – that is, no dissolved substance present.

Different forms of carbon

Carbon is a non-metallic element which exists in more than one solid structural form. These are graphite and diamond. Each of the forms has a different structure (Figures 3.28 and 3.29) and so they exhibit different physical properties (Table 3.2). The different physical properties that they exhibit lead to graphite and diamond being used in different ways (Table 3.3 and Figure 3.30).

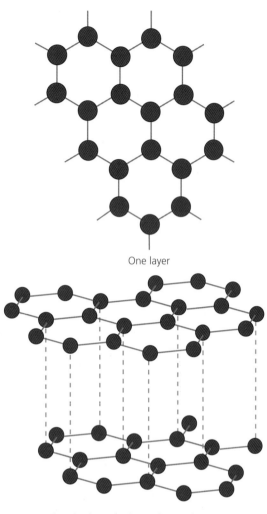

One layer

Showing how the layers fit together

a A portion of the graphite structure

b A piece of graphite as imaged through a scanning tunnelling microscope

▲ **Figure 3.28**

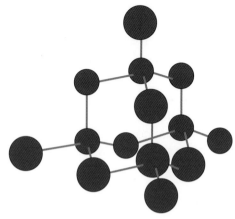

A small part of the structure

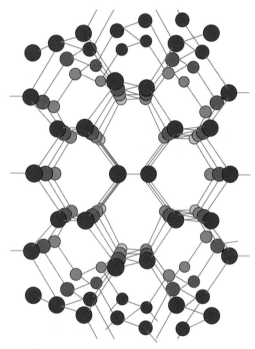

A view of a much larger part of the structure

a The structure of diamond

b The Regent Diamond has been worn by Queen Elizabeth II

▲ **Figure 3.29**

▼ **Table 3.2** Physical properties of graphite and diamond

Property	Graphite	Diamond
Appearance	A dark grey, shiny solid	A colourless transparent crystal which sparkles in light
Electrical conductivity	Conducts electricity	Does not conduct electricity
Hardness	A soft material with a slippery feel	A very hard substance
Density (g/cm³)	2.25	3.51

▼ **Table 3.3** Uses of graphite and diamond

Graphite	Diamond
Pencils	Jewellery
Electrodes	Glass cutters
Lubricant	Diamond-studded saws
	Drill bits
	Polishers

Graphite

Figure 3.28a shows the structure of graphite. This is a layer structure. Within each layer, each carbon atom is bonded to three others by strong covalent bonds. Each layer is therefore like a giant molecule. Between these layers there are weak forces of attraction and so the layers will pass over each other easily.

With only three covalent bonds formed between carbon atoms within the layers, an unbonded electron is present on each carbon atom. These 'spare' (or **delocalised**) electrons form electron clouds between the layers and it is because of these spare electrons that graphite conducts electricity.

Going further

Graphitic compounds

In recent years, a set of interesting compounds known as graphitic compounds have been developed. In these compounds, different atoms have been fitted in between the layers of carbon atoms to produce a substance with a greater electrical conductivity than pure graphite. Graphite is also used as a component in certain sports equipment, such as tennis and squash rackets.

Going further

Graphene

Discovered in 2004, graphene is a so-called super material made up of single layers of graphite as shown in the upper diagram of Figure 3.28a. It is able to conduct electricity one million times better than copper metal and has enormous potential in electronics.

Diamond

Figure 3.29a shows the diamond structure. Each of the carbon atoms in the giant structure is covalently bonded to four others. They form a tetrahedral arrangement similar to that found in silicon(IV) oxide (p. 44). This bonding scheme gives rise to a very rigid three-dimensional structure and accounts for the extreme hardness of the substances silicon(IV) oxide and diamond. All the outer shell electrons of the carbon atoms are used to form covalent bonds, so there are no electrons available to enable diamond or silicon(IV) oxide to conduct electricity.

It is possible to manufacture the different allotropes of carbon. Diamond is made by heating graphite to about 300°C at very high pressures. Diamond made by this method is known as industrial diamond. Graphite can be made by heating a mixture of coke and sand at a very high temperature in an electric arc furnace for about 24 hours.

The various uses of graphite and diamond result from their differing properties (Figure 3.30).

▲ **Figure 3.30** Uses of graphite (as a pencil 'lead' and in a squash racket) and diamond (as a toothed saw to cut marble and on a dentist's drill)

⊙ Going further

Buckminsterfullerene – an unusual form of carbon

In 1985, a new allotrope of carbon was obtained by Richard Smalley and Robert Curl of Rice University, Texas. It was formed by the action of a laser beam on a sample of graphite.

The structure of buckminsterfullerene can be seen in Figure 3.31.

This spherical structure is composed of 60 carbon atoms covalently bonded together. Further spherical forms of carbon, 'bucky balls', containing 70, 72 and 84 carbon atoms have been identified and the discovery has led to a whole new branch of inorganic carbon chemistry. It is thought that this type of molecule exists in chimney soot. Chemists have suggested that, due to the large surface area of the bucky balls, they may have uses as catalysts (Chapter 7, p. 106). Also they may have uses as superconductors.

Buckminsterfullerene is named after an American architect, Buckminster Fuller, who built complex geometrical structures (Figure 3.32).

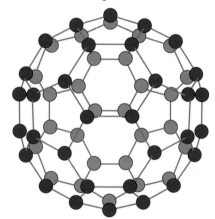

▲ **Figure 3.31** Buckminsterfullerene – a 'bucky ball' (C_{60})

▲ **Figure 3.32** C_{60} has a structure similar to a football and to the structure of the domes shown

Test yourself

10 Explain the difference between ionic and covalent bonding.

> **11** Compare the structures of silicon(IV) oxide and diamond.

12 Draw up a table to summarise the properties of the different types of substances you have met in this chapter. Your table should include examples from ionic substances and covalent substances (simple and giant).

Going further

Glasses and ceramics

Glasses

Glasses are all around us. Glasses are irregular giant molecular structures held together by strong covalent bonds. Glass can be made by heating silicon(IV) oxide with other substances until a thick viscous liquid is formed. As this liquid cools, the atoms present cannot move freely enough to return to their arrangement within the pure silicon(IV) oxide structure. Instead they are forced to form a disordered arrangement as shown in Figure 3.33. Glass is called a supercooled liquid.

• O ● Si

▲ **Figure 3.33** Two-dimensional structure of silicon(IV) oxide in glass

Ceramics

The word ceramic comes from the Greek word *keramos* meaning pottery or 'burnt stuff'. Clay dug from the ground contains a mixture of several materials including silicon(IV) oxide. During firing in a furnace, the clay is heated to a temperature of 1000°C. The material produced at the end of the firing, the ceramic, which is a type of glass, consists of many tiny mineral crystals bonded together with glass.

There are now many uses of these ceramic materials, such as ceramic discs, which prevent short-circuits in the national grid pylons; ceramic tiles, which protect the spacecraft; and ceramic bearings that do not need lubrication.

3.3 Metallic bonding

Another way in which atoms obtain a more stable electronic configuration is found in metals. The electrons in the outer shell of the atom of a metal move freely throughout the structure. They are delocalised, forming a mobile 'sea' of electrons (Figure 3.34). When the metal atoms lose these electrons, they form a giant lattice of positive ions. Therefore, metals consist of positive ions embedded in a sea of moving electrons. The negatively charged electrons attract all the positive metal ions and bond them together with strong electrostatic forces of attraction as a single unit. This is the **metallic bond**.

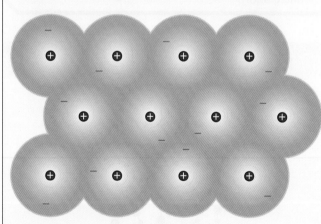

▲ **Figure 3.34** Metals consist of positive ions surrounded by a 'sea' of delocalised electrons

Properties of metals

Metals have the following properties.

» They conduct electricity due to the mobile electrons within the metal structure. When a metal is connected in a circuit, the electrons move towards the positive terminal while at the same time electrons are fed into the other end of the metal from the negative terminal.

» They are malleable and ductile. Unlike the fixed bonds in diamond, metallic bonds are not rigid, but they are still strong. If a force is applied to a metal, rows of ions can slide over one another. They reposition themselves and the strong bonds re-form as shown in Figure 3.35. 'Malleable' means that metals can be bent or hammered into different shapes. 'Ductile' means that the metals can be pulled out into thin wires.

» They usually have high melting and boiling points due to the strong attraction between the positive metal ions and the mobile 'sea' of electrons.

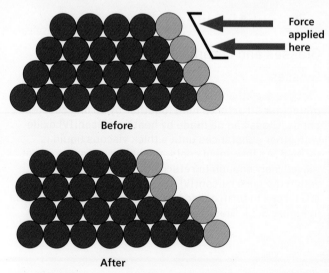

▲ **Figure 3.35** The positions of the positive ions in a metal before and after a force has been applied

» They have high densities because the atoms are very closely packed in a regular manner, as can be seen in Figure 3.36. Different metals have different types of packing of atoms and in doing so they produce the arrangements of ions shown in Figure 3.37.

▲ **Figure 3.36** Arrangement of ions in the crystal lattice of a metal

Going further

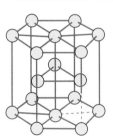

Close-packed
hexagonal structure
(e.g. magnesium,
density=
1.74 g cm^{-3})

Body-centred
cubic structure
(e.g. iron,
density=
7.87 g cm^{-3})

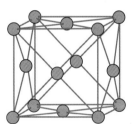

Face-centred
cubic structure
(e.g. copper,
density=
8.92 g cm^{-3})

▲ **Figure 3.37** Relating different structures to the density of metals

Revision checklist

After studying Chapter 3 you should be able to:

✔ Explain how compounds form between metals and non-metals.
✔ Describe the properties of ionic compounds.
✔ Explain in terms of structure and bonding the properties of ionic compounds.

> ✔ Work out the formulae of ionic substances from the charges on the ions.

✔ Identify redox reactions as reactions involving gain and loss of oxygen.

> ✔ Identify redox reactions as reactions involving gain and loss of electrons.
> ✔ Identify redox reactions by changes in oxidation number.

✔ Define redox reactions as involving simultaneous oxidation and reduction.
✔ Identify oxidation as gain of oxygen and reduction as loss of oxygen.

> ✔ Define oxidation as loss of electrons and increase in oxidation number, and reduction as gain of electrons and reduction in oxidation number.
> ✔ Identify redox reactions by the colour changes involved when using acidified potassium manganate(VII) or potassium iodide.
> ✔ Define an oxidising agent as a substance that oxidises another substance and is itself reduced during a redox reaction.
> ✔ Define a reducing agent as a substance that reduces another substance and is itself oxidised during a redox reaction.

✔ Identify oxidation and reduction in redox reactions.

> ✔ Identify oxidising agents and reducing agents in redox reactions.

✔ Explain how simple compounds form between non-metals by single covalent bonds.

> ✔ Explain how simple compounds containing multiple covalent bonds form between non-metals.

✔ Describe and explain the properties of simple covalent compounds.
✔ Describe the giant covalent structures of graphite and diamond.

> ✔ Describe the giant covalent structure of silicon(IV) oxide, SiO_2.

✔ Relate the structures of graphite and diamond to their uses.

> ✔ Describe the similarity in properties between diamond and silicon(IV) oxide, related to their structures.
> ✔ Describe the arrangement of particles in a solid metal.
> ✔ Recognise that the particles in a metal are held together by metallic bonds.
> ✔ Describe metallic bonding.
> ✔ Use metallic bonding to explain some of the properties of metals.

Exam-style questions

1 Define the following:
   a positive (+) ion [2]
   b negative (–) ion [2]
   c ionic bond [2]
   d electrostatic force of attraction. [2]
2 Using Table 3.1 on p. 35, write down the names of:
   a three atoms which would form an ion
   with a charge of +2 [3]
   b three atoms which would form an ion
   with a charge of +1 [3]
   c three atoms/groups of atoms which would
   form an ion with a charge of –2. [3]

---

3 Sketch diagrams to show the bonding in each
   of the following compounds:
   a calcium fluoride (CaF_2) [4]
   b oxygen (O_2) [4]
   c magnesium chloride ($MgCl_2$) [4]
   d tetrachloromethane (CCl_4). [4]
4 Use the information given in Table 3.1
   on p. 35 to deduce the formula for:
   a silver oxide
   b zinc chloride
   c potassium sulfate
   d calcium nitrate
   e iron(II) nitrate
   f copper(II) carbonate
   g iron(III) hydroxide
   h aluminium fluoride. [8]

---

5 Atoms of elements X, Y and Z have 16, 17 and
   19 electrons, respectively. Atoms of argon have
   18 electrons.
   a Determine the formulae of the compounds
   formed by the combination of the atoms of
   the elements:
      i X and Z [1]
      ii Y and Z [1]
      iii X with itself. [1]
   b In each of the cases shown in **a i–iii** above,
   identify the type of chemical bond formed. [3]
   c Give two properties you would expect each
   of the compounds formed in **a ii** and **iii**
   to have. [4]

---

6 Explain the following statements.
   a Ammonia is a gas at room temperature. [3]
   b The melting points of sodium chloride
   and iodine are very different. [8]
   c Metals generally are good conductors
   of electricity. [3]
   d Graphite acts as a lubricant but diamond
   does not. [3]

---

4 Stoichiometry – chemical calculations

In this chapter, you will learn how to calculate the masses of products formed in a chemical reaction, when told the quantity of reactants you start with, and also how to calculate the mass of starting materials you need to produce a certain mass of product. These are the type of calculations your teacher does before they write the methods for student practicals. Have you ever thought how your teachers were able to get it right every time?

We will look at the amounts of substances used and formed during reactions involving solids, gases and solutions. You need to know whether you actually got the right amount of product in your experiment, and we will also look at how well a reaction went.

By the end of this chapter we hope that you will be confident in using numbers and balanced chemical reactions to help you to discover these things.

4.1 Relative atomic mass

There are at present 118 different elements known. The atoms of these elements differ in mass because of the different numbers of protons, neutrons and electrons they contain. The actual mass of one atom is very small. For example, the mass of a single atom of sulfur is around:

0.000 000 000 000 000 000 000 053 16 g

Such small quantities are not easy to work with and, as you saw in Chapter 3, a scale called the **relative atomic mass**, A_r, scale is used. In this scale, an atom of carbon is given a relative atomic mass, A_r, of 12.00. All other atoms of the other elements are given a relative atomic mass compared to that of carbon.

An H atom is $\frac{1}{12}$ the mass of a C atom | An Mg atom is twice the mass of a C atom

H	C	Mg	S	Ca
1	12	24	32	40
	Fixed			

▲ **Figure 4.1** The relative atomic masses of the elements H, C, Mg, S and Ca

> **Key definition**
> **Relative atomic mass, A_r,** is the average mass of the isotopes of an element compared to 1/12th of the mass of an atom of ^{12}C.

Reacting masses

Chemists often need to be able to show the relative masses of the atoms involved in a chemical process.

❓ Worked example

What mass of carbon dioxide would be produced if 6 g of carbon was completely burned in oxygen gas?

$$C + O_2 \rightarrow CO_2$$

Instead of using the actual masses of atoms, we use the relative atomic mass to help us answer this type of question.

In this example we can work out the **relative molecular mass**, M_r, of molecules such as O_2 and CO_2 using the relative atomic masses of the atoms they are made from. The relative molecular mass is the sum of the relative atomic masses of all those elements shown in the **molecular formula** of the substance. The molecular formula of oxygen is O_2 and it shows that one molecule of oxygen contains two oxygen atoms. Each oxygen atom has a relative atomic mass of 16, so O_2 has a relative molecular mass of 2 × 16 = 32.

The molecular formula of carbon dioxide is CO_2 and it shows that in one molecule of carbon dioxide there is one carbon atom and two oxygen atoms. The carbon atom has a relative atomic mass of 12 and each oxygen atom has a relative atomic mass of 16, so CO_2 has a relative molecular mass of $12 + (2 \times 16) = 44$.

> **Key definitions**
>
> The **molecular formula** of a compound is defined as the number and type of different atoms in one molecule.
>
> **Relative molecular mass**, M_r, is the sum of the relative atomic masses. Relative formula mass, M_r, is used for ionic compounds.
>
> **Relative formula mass**, M_r, is used for ionic compounds

So we can now use the equation to answer the question: What mass of carbon dioxide would be produced if 6 g of carbon was completely combusted?

$$C + O_2 \rightarrow CO_2$$
$$12 \quad 32 \quad\;\; 44$$

We can convert these relative masses to actual masses by adding mass units, g, which gives:

$$C + O_2 \rightarrow CO_2$$
$$12 \quad 32 \quad\;\; 44$$
$$12\,g \quad 32\,g \quad 44\,g$$

The above calculation shows that if 12 g of carbon was burned completely then 44 g of carbon dioxide gas would be formed. So, 6 g of carbon burning would result in the formation of 22 g of carbon dioxide gas.

> ▶ **Test yourself**
>
> 1 What mass of carbon dioxide gas would be produced if 10 g of calcium carbonate reacted with an excess of hydrochloric acid?
> 2 What mass of sulfur dioxide would be produced if 64 tonnes of sulfur was completely reacted with oxygen gas?

Chemists often need to know how much of a substance has been formed or used up during a chemical reaction. This is particularly important in the chemical industry, where the substances being reacted (the reactants) and the substances being produced (the products) may be worth a great deal of money. Waste costs money!

To solve this problem, the chemical industry needs a way of counting atoms, ions or molecules. As atoms, ions and molecules are very tiny particles, it is impossible to measure out a dozen or even a hundred of them. Instead, chemists weigh out a very large number of particles.

> This number is 6.02×10^{23} atoms, ions or molecules and is called Avogadro's constant after the famous Italian scientist Amedeo Avogadro (1776–1856). An amount of substance containing 6.02×10^{23} particles is called a **mole** (often abbreviated to mol).

> **Key definition**
>
> The **mole**, symbol mol, is the unit of amount of substance. 1 mole contains 6.02×10^{23} particles, e.g. atoms, ions, molecules. This number is called the Avogadro constant.

4.2 Calculating moles

In Chapter 3, we looked at how we can compare the masses of all the other atoms with the mass of carbon atoms. This is the basis of the relative atomic mass scale. Chemists have found by experiment that if you take the relative atomic mass of an element in grams, it always contains 6.02×10^{23} or 1 mole of its atoms.

Moles and elements

For the elements we can see that, for example, the relative atomic mass (A_r) of iron is 56, so 1 mole of iron is 56 g. Therefore, 56 g of iron contains 6.02×10^{23} atoms.

The A_r for aluminium is 27. In 27 g of aluminium it is found that there are 6.02×10^{23} atoms. Therefore, 27 g of aluminium is 1 mole of aluminium atoms.

So, we can calculate the mass of a substance present in any number of moles using the relationship:

mass (in grams) = number of moles × molar mass
of the element

The molar mass of an element or compound is the mass of 1 mole of the element or compound and has units of g/mol.

So, a mole of the element magnesium is 6.02×10^{23} atoms of magnesium and a mole of the element carbon is 6.02×10^{23} atoms of carbon (Figure 4.2).

a   A mole of magnesium

b   A mole of carbon

▲ **Figure 4.2**

? **Worked example**

Calculate the number of atoms of carbon in:

a   0.5 moles

b   0.1 moles.

Molar mass of carbon contains 6.02×10^{23} atoms of carbon so 0.5 moles would contain half of this amount:

Atoms of carbon in 0.5 moles = $6.02 \times 10^{23} \times 0.5 = 3.01 \times 10^{23}$ atoms

In a similar way:

Atoms of carbon in 0.1 moles = $6.02 \times 10^{23} \times 0.1 = 6.02 \times 10^{22}$ atoms

? **Worked example**

What mass of hydrogen gas would be produced if 46 g of sodium was reacted with water?

First write down the balanced chemical equation:

$$2Na + 2H_2O \rightarrow 2NaOH + H_2$$

Next find the relative atomic mass of sodium (from the Periodic Table (p. 135)) and work out the relative formula masses of water, sodium hydroxide and hydrogen gas. The term **relative formula mass** is used when the compound is ionic, in this case sodium hydroxide. It is used and calculated in the same way as we have used relative molecular mass for covalently bonded compounds, such as water and hydrogen.

Relative atomic mass of sodium is 23.

Relative molecular mass of water, H_2O, is $(2 \times 1) + 16 = 18$

Relative formula mass of sodium hydroxide is $23 + 16 + 1 = 40$

Relative molecular mass of hydrogen gas, H_2, is $2 \times 1 = 2$

Now write these masses under the balanced chemical equation taking into account the numbers used to balance the equation.

$2Na$	+	$2H_2O$	→	$2NaOH$	+	H_2
$2 \times 23 = 46$		$2 \times 18 = 36$		$2 \times 40 = 80$		2

These relative masses can now be converted into actual or reacting masses by putting in mass units, for example, grams.

$2Na$	+	$2H_2O$	→	$2NaOH$	+	H_2
$2 \times 23 = 46$		$2 \times 18 = 36$		$2 \times 40 = 80$		2
46 g		36 g	→	18 g		2g

So the answer to the question of what mass of hydrogen would be produced if 46 g of sodium was reacted with water is 2 g.

Worked example

Calculate the mass of **a** 2 moles of iron and **b** 0.25 mole of iron. (A_r: Fe = 56)

a   mass of 2 moles of iron

= number of moles × relative atomic mass (A_r)

= 2 × 56

= 112 g

b   mass of 0.25 mole of iron

= number of moles × relative atomic mass (A_r)

= 0.25 × 56

= 14 g

If we know the mass of the element, then it is possible to calculate the number of moles of that element using:

$$\text{number of moles} = \frac{\text{mass of the element}}{\text{molar mass}}$$

 Worked example

Calculate the number of moles of aluminium present in **a** 108 g and **b** 13.5 g of the element. (A_r: Al = 27)

a   number of moles of aluminium

$= \dfrac{\text{mass of aluminium}}{\text{molar mass of aluminium}}$

$= \dfrac{108}{27}$

= 4 moles

b   number of moles of aluminium

$= \dfrac{\text{mass of aluminium}}{\text{molar mass of aluminium}}$

$= \dfrac{13.5}{27}$

= 0.5 mole

4.3 Moles and compounds

The idea of the mole can also be used with compounds (Figure 4.3).

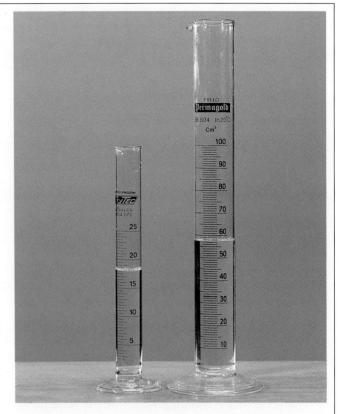

▲ **Figure 4.3** 1 mole of water (H_2O) (left) and 1 mole of ethanol (C_2H_5OH) (right) in separate measuring cylinders

For example, consider the molar mass of water (H_2O) molecules (A_r: H = 1; O = 16).

From the formula of water, H_2O, you will see that 1 mole of water molecules contains 2 moles of hydrogen (H) atoms and 1 mole of oxygen (O) atoms. The molar mass of water molecules is therefore:

(2 × 1) + (1 × 16) = 18 g

The molar mass of a compound is called its molar mass: it has units of g/mol. If you write the molar mass of a compound without any units then it is the relative molecular mass (M_r). So the relative molecular mass of water is 18.

Now follow these examples to help you understand more about moles and compounds.

What is **a** the molar mass and **b** the relative molecular mass, M_r, of ethanol, C_2H_5OH? (A_r: H = 1; C = 12; O = 16)

a   1 mole of C_2H_5OH contains 2 moles of carbon atoms, 6 moles of hydrogen atoms and 1 mole of oxygen atoms. Therefore:

$$\text{molar mass of ethanol} = (2 \times 12) + (6 \times 1) + (1 \times 16)$$
$$= 46 \text{ g/mol}$$

b   The relative molecular mass of ethanol is 46.

? **Worked example**

What is **a** the molar mass and **b** the relative molecular mass of nitrogen gas, N_2? (A_r: N = 14)

a   Nitrogen is a diatomic gas. Each nitrogen molecule contains two atoms of nitrogen. Therefore:
molar mass of $N_2 = 2 \times 14$
$$= 28 \text{ g/mol}$$

b   The relative molecular mass of N_2 is 28.

The mass of a compound found in any number of moles can be calculated using the relationship:

$$\text{mass of compound} = \text{number of moles of the compound} \times \text{molar mass of the compound}$$

? **Worked example**

Calculate the mass of **a** 3 moles and **b** 0.2 moles of carbon dioxide gas, CO_2. (A_r: C = 12; O = 16)

a   1 mole of CO_2 contains 1 mole of carbon atoms and 2 moles of oxygen atoms. Therefore:
molar mass of $CO_2 = (1 \times 12) + (2 \times 16)$
$$= 44 \text{ /mol}$$
mass of 3 moles = number of × molar mass
of CO_2 moles   of CO_2
$$= 3 \times 44$$
$$= 132 \text{ g}$$

b   mass of 0.2 mole = number of × molar mass
of CO_2 moles   of CO_2
$$= 0.2 \times 44$$
$$= 8.8 \text{ g}$$

If we know the mass of the compound, then we can calculate the number of moles of the compound using the relationship:

$$\text{number of moles of compound} = \frac{\text{mass of compound}}{\text{molar mass of the compound}}$$

? **Worked example**

Calculate the number of moles of magnesium oxide, MgO, in **a** 80 g and **b** 10 g of the compound. (A_r: O = 16; Mg = 24)

a   1 mole of MgO contains 1 mole of magnesium atoms and 1 mole of oxygen atoms. Therefore:
molar mass of MgO = (1 × 24) + (1 × 16) = 40 g/mol
number of moles of MgO in 80 g

$$= \frac{\text{mass of MgO}}{\text{molar mass of MgO}} = \frac{80}{40}$$

= 2 moles

b   number of moles of MgO in 10 g

$$= \frac{\text{mass of MgO}}{\text{molar mass of MgO}} = \frac{10}{40}$$

= 0.25 mole

Moles and gases

Many substances exist as gases. If we want to find the number of moles of a gas, we can do this by measuring the volume rather than the mass.

Chemists have shown by experiment that 1 mole of any gas occupies a volume of approximately 24 dm³ (24 litres) at room temperature and pressure (r.t.p.). This quantity is also known as the molar gas volume, V_m.

Therefore, it is relatively easy to convert volumes of gases into moles and moles of gases into volumes using the following relationship:

$$\text{number of moles of a gas} = \frac{\text{volume of the gas (in dm}^3 \text{ at r.t.p)}}{24 \text{ dm}^3}$$

or

$$\text{volume of a gas (in dm}^3 \text{ at r.t.p.)} = \text{number of moles of gas} \times 24 \text{ dm}^3$$

We will use these two relationships to help us answer some questions concerning gases.

? **Worked example**

Calculate the number of moles of ammonia gas, NH_3, in a volume of 72 dm³ of the gas measured at r.t.p.

$$\text{number of moles of ammonia} = \frac{\text{volume of ammonia in dm}^3}{24 \text{ dm}^3}$$

$$= \frac{72}{24} = 3$$

> **?** **Worked example**
>
> Calculate the volume of carbon dioxide gas, CO_2, occupied by **a** 5 moles and **b** 0.5 mole of the gas measured at r.t.p.
>
> a   volume of CO_2 = number of moles of $CO_2 \times 24$ dm^3
> $= 5 \times 24 = 120$ dm^3
>
> b   volume of CO_2 = number of moles of $CO_2 \times 24$ dm^3
> $= 0.5 \times 24 = 12$ dm^3

The volume occupied by 1 mole of any gas must contain 6.02×10^{23} molecules. Therefore, it follows that equal volumes of all gases measured at the same temperature and pressure must contain the same number of molecules. This hypothesis was also first put forward by Amedeo Avogadro and is called **Avogadro's Law.**

Moles and solutions

Chemists often need to know the concentration of a solution. Sometimes **concentration** is measured in grams per cubic decimetre (g/dm^3), but more often concentration is measured in moles per cubic decimetre (mol/dm^3).

> **Key definition**
>
> **Concentration** can be measured in g/dm^3 or mol/dm^3.

When 1 mole of a substance is dissolved in water and the solution is made up to 1 dm^3 (1000 cm^3), a 1 molar (1 mol/dm^3) solution is produced. Chemists do not always need to make up such large volumes of solution.

A simple method of calculating the concentration uses the relationship:

$$\text{concentration (in } mol/dm^3) = \frac{\text{number of moles}}{\text{volume (in } dm^3)}$$

It is very easy to change a volume given in cm^3 into one in dm^3 by simply dividing the volume in cm^3 by 1000. For example, 250 cm^3 is 0.25 dm^3.

> **?** **Worked example**
>
> Calculate the concentration (in mol/dm^3) of a solution of sodium hydroxide, NaOH, which was made by dissolving 10 g of solid sodium hydroxide in water and making up to 250 cm^3. (A_r: H = 1; O = 16; Na = 23)
>
> 1 mole of NaOH contains 1 mole of sodium, 1 mole of oxygen and 1 mole of hydrogen. Therefore:
>
> molar mass of NaOH = $(1 \times 23) + (1 \times 16) + (1 \times 1)$
> $= 40$ g
>
> $$\frac{\text{number of moles}}{\text{NaOH in 10 g}} = \frac{\text{mass of NaOH}}{\text{molar mass of NaOH}}$$
>
> $$= \frac{10}{40} = 0.25$$
>
> $$(250 \text{ } cm^3 = \frac{250}{1000} \text{ } dm^3 = 0.25 \text{ } dm^3)$$
>
> $$\frac{\text{concentration of the}}{\text{NaOH solution}} = \frac{\text{number of moles of NaOH}}{\text{volume of solution (} dm^3)}$$
>
> $$= \frac{0.25}{0.25} = 1 \text{ } mol/dm^3$$

Sometimes chemists need to know the mass of a substance that has to be dissolved to prepare a known volume of solution at a given concentration. A simple method of calculating the number of moles, and so the mass of substance needed, is by using the relationship:

$$\begin{array}{ccc}\text{number of} & = \text{concentration} & \times \text{volume in solution} \\ \text{moles} & \text{(in } mol/dm^3) & \text{(in } dm^3) \end{array}$$

> **?** **Worked example**
>
> Calculate the mass of potassium hydroxide, KOH, that needs to be used to prepare 500 cm^3 of a 2 mol/dm^3 solution in water. (A_r: H = 1; O = 16; K = 39)
>
> number of moles of KOH
>
> $\begin{array}{cc} = \text{concentration of solution} & \times \text{volume of solution} \\ \text{(} mol/dm^3) & \text{(} dm^3) \end{array}$
>
> $$= 2 \times \frac{500}{1000} = 1$$
>
> 1 mole of KOH contains 1 mole of potassium, 1 mole of oxygen and 1 mole of hydrogen. Therefore:
>
> molar mass of KOH
> $= (1 \times 39) + (1 \times 16) + (1 \times 1)$
> $= 56$ g
>
> Therefore:
> mass of KOH in 1 mole (500 cm^3 of 2 mol/dm^3) = 56 g

> ### Test yourself
>
> Use these values of A_r to answer the questions below.
> H = 1; C = 12; N = 14; O = 16; Ne = 20; Na = 23; Mg = 24; S = 32; Cl = 35.5; K = 39; Fe = 56; Cu = 63.5; Zn = 65
> 1 mole of any gas at r.t.p. occupies 24 dm³.
>
> **3** Calculate the number of moles in:
>   **a** 2 g of neon atoms
>   **b** 4 g of magnesium atoms
>   **c** 24 g of carbon atoms.
> **4** Calculate the number of atoms in 1 mole of:
>   **a** calcium
>   **b** carbon dioxide
>   **c** methane.
> **5** Calculate the mass of:
>   **a** 0.1 mol of oxygen molecules
>   **b** 5 mol of sulfur atoms
>   **c** 0.25 mol of sodium atoms.
> **6** Calculate the number of moles in:
>   **a** 9.8 g of sulfuric acid (H_2SO_4)
>   **b** 40 g of sodium hydroxide (NaOH)
>   **c** 720 g of iron(II) oxide (FeO).
> **7** Calculate the mass of:
>   **a** 2 mol of zinc oxide (ZnO)
>   **b** 0.25 mol of hydrogen sulfide (H_2S)
>   **c** 0.35 mol of copper(II) sulfate ($CuSO_4$).
> **8** Calculate the number of moles at r.t.p. in:
>   **a** 2 dm³ of carbon dioxide (CO_2)
>   **b** 240 dm³ of sulfur dioxide (SO_2)
>   **c** 20 cm³ of carbon monoxide (CO).
> **9** Calculate the volume of:
>   **a** 0.3 mol of hydrogen chloride (HCl)
>   **b** 4.4 g of carbon dioxide
>   **c** 34 g of ammonia (NH_3).
> **10** Calculate the concentration of solutions containing:
>   **a** 0.2 mol of sodium hydroxide dissolved in water and made up to 100 cm³
>   **b** 9.8 g of sulfuric acid dissolved in water and made up to 500 cm³.
> **11** Calculate the mass of:
>   **a** copper(II) sulfate ($CuSO_4$) which needs to be used to prepare 500 cm³ of a 0.1 mol/dm³ solution
>   **b** potassium nitrate (KNO_3) which needs to be used to prepare 200 cm³ of a 2 mol/dm³ solution.
> **12** How many atoms are there in:
>   **a** 0.25 moles of aluminium
>   **b** 0.15 moles of magnesium?
> **13** How many moles of calcium contain 1.204×10^{24} atoms?

4.4 Calculating formulae

If we have 1 mole of a compound, then the formula shows the number of moles of each element in that compound. For example, the formula for lead(II) bromide is $PbBr_2$. This means that 1 mole of lead(II) bromide contains 1 mole of lead ions and 2 moles of bromide ions. If we do not know the formula of a compound, we can find the masses of the elements present experimentally, and we can then use these masses to work out the formula of that compound.

> ### Practical skills
>
> Experiment to find the formula of magnesium oxide
>
> For safe experiments/demonstrations which are related to this chapter, please refer to the *Cambridge IGCSE Chemistry Practical Skills Workbook*, which is also part of this series.
>
> Safety
> ● Eye protection must be worn.
> ● Take care when handling hot apparatus.
>
>
>
> ▲ **Figure 4.4** Apparatus used to determine magnesium oxide's formula
>
> If a known mass of magnesium ribbon is heated strongly in a crucible (or 'bottle-top' crucible) with a lid, it reacts with oxygen to form magnesium oxide.

magnesium + oxygen → magnesium oxide

To allow oxygen into the crucible, the lid needs to be lifted briefly during the heating process.

To heat to constant mass:

- Heat strongly for 5 minutes, then the crucible and contents should be allowed to cool and reweighed.
- The above step should be repeated until a constant mass is obtained.

When the magnesium is heated, it can be seen that:

- the shiny magnesium metal burns brightly to form a white powder, magnesium oxide
- the powder is very fine and when the lid is lifted it can be seen to rise out of the crucible (which is the reason the lid is lifted only briefly).

▼ **Table 4.1** Data from the magnesium oxide experiment

Mass of crucible	14.63 g
Mass of crucible and magnesium	14.87 g
Mass of crucible and magnesium oxide	15.03 g

From these data we can find the formula of magnesium oxide by following these steps:

1 Calculate the mass of magnesium metal used in this experiment.
2 Calculate the mass of oxygen which reacts with the magnesium.
3 Calculate the number of moles of magnesium used.
4 Calculate the number of moles of oxygen atoms which react with the magnesium.
5 Determine the simplest ratio of moles of magnesium to moles of oxygen.
6 Determine the simplest formula for magnesium oxide.

This formula is the **empirical formula** of the compound.

> **Key definition**
>
> The **empirical formula** of a compound is the simplest whole number ratio of the different atoms or ions in a compound.

7 Why is it important to lift the lid briefly while heating the crucible?
8 What are the main sources of error in this experiment?

❓ Worked example

In an experiment, an unknown organic compound was found to contain 0.12 g of carbon and 0.02 g of hydrogen. Calculate the empirical formula of the compound. (A_r: H = 1; C = 12)

	C	**H**
Masses (g)	0.12	0.02
Number of moles	$\frac{0.12}{12} = 0.01$	$\frac{0.02}{1} = 0.02$
Ratio of moles	1	2
Empirical formula	CH_2	

From our knowledge of covalent bonding (Chapter 3, p. 38) we know that a molecule of this formula cannot exist. However, molecules with the following formulae do exist: C_2H_4, C_3H_6, C_4H_8 and C_5H_{10}. All of these formulae show the same ratio of carbon atoms to hydrogen atoms, CH_2, as our unknown. To find out which of these formulae is the actual formula for the unknown organic compound, we need to know the molar mass of the compound.

Using a mass spectrometer, the relative molecular mass (M_r) of this organic compound was found to be 56. We need to find out the number of empirical formulae units present:

M_r of the empirical formula unit

$= (1 \times 12) + (2 \times 1)$

$= 14$

Number of empirical formula units present

$= \dfrac{M_r \text{ of compound}}{M_r \text{ of empirical formula unit}} = \dfrac{56}{14}$

$= 4$

Therefore, the actual formula of the unknown organic compound is $4 \times CH_2 = C_4H_8$.

This substance is called butene. C_4H_8 is the **molecular formula** for this substance and shows the actual numbers of atoms of each element present in one molecule of the substance.

> **Key definition**
> The **molecular formula** of a compound is the number and type of different atoms in one molecule.

Sometimes the composition of a compound is given as a percentage by mass of the elements present. In cases such as this, the procedure shown in the next example is followed.

? Worked example

Calculate the empirical formula of an organic compound containing 92.3% carbon and 7.7% hydrogen by mass. The M_r of the organic compound is 78. What is its molecular formula? (A_r: H = 1; C = 12)

	C	H
% by mass	92.3	7.7
Masses in 100 g	92.3 g	7.7 g
Number of moles	$\frac{92.3}{12} = 7.7$	$\frac{7.7}{1} = 7.7$
Ratio of moles	1	1
Empirical formula	CH	

M_r of the empirical formula unit CH

$= 12 + 1$

$= 13$

Number of empirical formula units present

$= \dfrac{M_r \text{ of compound}}{M_r \text{ of empirical formula unit}} = \dfrac{78}{13}$

$= 6$

The molecular formula of the organic compound is $6 \times CH = C_6H_6$. This is a substance called benzene.

▶ Test yourself

Use the following values of A_r to answer the questions below: H = 1; C = 12; O = 16; Ca = 40.
14 Determine the empirical formula of an oxide of calcium formed when 0.4 g of calcium reacts with 0.16 g of oxygen.
15 Determine the empirical formula of an organic hydrocarbon compound which contains 80% by mass of carbon and 20% by mass of hydrogen. If the M_r of the compound is 30, what is its molecular formula?

4.5 Moles and chemical equations

When we write a balanced chemical equation, we are indicating the numbers of moles of reactants and products involved in the chemical reaction. Consider the reaction between magnesium and oxygen.

$$\text{magnesium} + \text{oxygen} \rightarrow \text{magnesium oxide}$$

$$2Mg(s) + O_2(g) \rightarrow 2MgO(s)$$

This shows that 2 moles of magnesium react with 1 mole of oxygen to give 2 moles of magnesium oxide.

Using the ideas of moles and masses we can use this information to calculate the quantities of the different chemicals involved.

$2Mg(s)$	+	$O_2(g)$	→	$2MgO(s)$
2 moles		1 mole		2 moles
2×24		$1 \times (16 \times 2)$		$2 \times (24 + 16)$
$= 48$ g		$= 32$ g		$= 80$ g

You will notice that the total mass of reactants is equal to the total mass of product. This is true for any chemical reaction and it is known as the **Law of conservation of mass**. This law was understood by the Greeks but was first clearly formulated by Antoine Lavoisier in 1774. Chemists can use this idea to calculate masses of products formed and reactants used in chemical processes before they are carried out.

Solids

Lime (calcium oxide, CaO) is used in the manufacture of lime mortar: a mixture of lime, sand, aggregate and water. It is used as a binding material when building with brick. Lime is manufactured in large quantities in Europe (Figure 4.5) by heating limestone (calcium carbonate, $CaCO_3$).

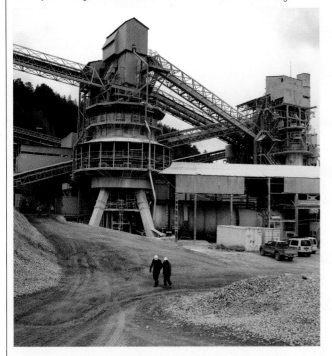

▲ **Figure 4.5** A rotary kiln for burning (calcining) limestone into lime, located in Moha, Belgium

The equation for the process is:

$$CaCO_3(s) \rightarrow CaO(s) + CO_2(g)$$

1 mole	1 mole	1 mole
$40 + 12 + (3 \times 16)$	$40 + 16$	$12 + (2 \times 16)$
$= 100\,g$	$= 56\,g$	$= 44\,g$

Calculate the amount of lime produced when 10 tonnes of limestone are heated (Figure 4.6). (A_r: C = 12; O = 16; Ca = 40)

1 tonne (t) = 1000 kg

1 kg = 1000 g

From this relationship between grams and tonnes we can replace the masses in grams by masses in tonnes:

$$CaCO_3(s) \rightarrow CaO(s) + CO_2(g)$$

100 t	56 t	44 t
10 t	5.6 t	4.4 t

10 tonnes of limestone

Heat

? tonnes of lime

▲ **Figure 4.6** How much lime is produced?

The equation now shows that 100 t of limestone will produce 56 t of lime. Therefore, 10 t of limestone will produce 5.6 t of lime.

Gases

Many chemical processes involve gases. The volume of a gas is measured more easily than its mass. This example shows how chemists work out the volumes of gaseous reactants and products needed using Avogadro's Law and the idea of moles.

Worked example

Some rockets use hydrogen gas as a fuel. When hydrogen burns in oxygen it forms water vapour. Calculate the volumes of **a** $O_2(g)$ used and **b** water, $H_2O(g)$, produced if 960 dm^3 of hydrogen gas, $H_2(g)$, was burned in oxygen. (A_r: H = 1; O = 16) Assume 1 mole of any gas occupies a volume of 24 dm^3.

$$2H_2(g) \quad + \quad O_2(g) \quad \rightarrow \quad 2H_2O(g)$$

2 moles	1 mole	2 moles
2 × 24	1 × 24	2 × 24
= 48 dm^3	= 24 dm^3	= 48 dm^3

When 960 dm^3 of hydrogen is burned in oxygen:
a   480 dm^3 of oxygen is required and
b   960 dm^3 of $H_2O(g)$ is produced.

Therefore:

(×2)	96 dm^3	48 dm^3	96 dm^3
(×10)	960 dm^3	480 dm^3	960 dm^3

Solutions

Worked example

Chemists usually carry out reactions using solutions. If they know the concentration of the solution(s) they are using, they can find out the quantities reacting.

Calculate the volume of 1 mol/dm^3 solution of H_2SO_4 required to react completely with 6 g of magnesium. (A_r: Mg = 24).

$$\text{number of moles of magnesium}$$
$$= \frac{\text{mass of magnesium}}{\text{molar mass of magnesium}} = \frac{6}{24}$$
$$= 0.25$$

$$Mg(s) \quad + \quad H_2SO_4(aq) \quad \rightarrow \quad MgSO_4(aq) \quad + \quad H_2(g)$$

1 mole	1 mole	1 mole	1 mole
0.25 mol	0.25 mol	0.25 mol	0.25 mol

So 0.25 mol of $H_2SO_4(aq)$ is required. Using:

volume of $H_2SO_4(aq)$ (dm^3)

$$= \frac{\text{moles of } H_2SO_4}{\text{concentration of } H_2SO_4 \ (\text{mol/dm}^3)} = \frac{0.25}{1}$$

$$= 0.25 \ \text{dm}^3 \text{ or } 250 \ \text{cm}^3$$

Worked example

40 cm^3 of 0.2 mol/dm^3 solution of hydrochloric acid just neutralised 20 cm^3 of sodium hydroxide solution in a titration (Chapter 8, p. 128).

What is the concentration of sodium hydroxide solution in g/dm^3 used in this neutralisation reaction?

number of moles of HCl used
= concentration (mol/dm^3) × volume (dm^3) = 0.2 × 0.04
= 0.008

$$HCl(aq) \quad + \quad NaOH(aq) \quad \rightarrow \quad NaCl(aq) \quad + \quad H_2O(l)$$

1 mole	1 mole	1 mole	1 mole
0.008 mol	0.008 mol	0.008 mol	0.008 mol

You will see that 0.008 mole of NaOH was present. The concentration of the NaOH(aq) is given by:

concentration of NaOH (mol/dm3)

$$= \frac{\text{number of moles of NaOH}}{\text{volume of NaOH (dm}^3)} = \frac{0.008}{0.02}$$

(volume of NaOH in dm^3 = $\frac{20}{1000}$ = 0.02)

$$= 0.4 \ \text{mol/dm}^3$$

Now we have the concentration in mol/dm^3, we can easily convert this in g/dm^3 using the relationship below:

Concentration of a = concentration in × molar mass of
solution in g/dm^3 mol/dm^3 the substance

So the concentration of the NaOH in g/dm^3 = 0.4 × (23 + 16 + 1)
$$= 16 \text{g/dm}^3$$

Percentage yield

Chemical reactions rarely produce the predicted amount of product from the masses of reactants in the reaction as they are not 100% efficient.

An example of this is the reaction of carbon with oxygen to produce carbon dioxide gas.

$$C(s) + O_2(g) \rightarrow CO_2(g)$$

The equation for the reaction states that 1 mole of carbon reacts with oxygen to give 1 mole of carbon dioxide gas.

If you burn 12 g, 1 mole, of carbon to make CO_2, then the amount of carbon dioxide expected is 44 g, 1 mole of CO_2. The theoretical yield of carbon dioxide from this reaction is 44 g. This only occurs, however, if the reaction is 100% efficient.

In reality, the mass of carbon dioxide you will get will be less than 44 g, because another reaction can also occur between carbon and oxygen. Some of the carbon reacts to make carbon monoxide, CO.

$$2C(s) + O_2(g) \rightarrow 2CO(g)$$

The **percentage yield** of the reaction is based on the amount of carbon dioxide that is actually produced against what should have been produced if the reaction were 100% efficient.

For example, if 12 g of carbon was burned in excess oxygen and only 28 g of carbon dioxide was produced, the percentage yield can be worked out:

$$\text{percentage yield of carbon dioxide} = \frac{28}{44} \times 100$$

$$= 63.6\%$$

In general:

$$\text{percentage yield} = \frac{\text{actual yield}}{\text{theoretical yield}} \times 100$$

Percentage composition

Percent composition is used to describe the percent by mass of each element in a compound. It is found by dividing the mass of a particular element in the compound by the molar mass and then multiplying by 100 to give a percentage.

For example, the percentage composition of magnesium oxide, MgO, can be found by using the calculations shown below.

The molar mass of MgO is 24 + 16 = 40 g

Of this, 24g is magnesium and 16g is oxygen.

$$\%Mg = \frac{24}{40} \times 100 = 60\%$$

$$\%O = \frac{16}{40} \times 100 = 40\%$$

Percentage purity

In Chapter 2 (p. 14), we saw that the **purity** of a substance is very important. If a factory makes medicines or chemicals used in food then the purity of the product is crucial, as the impurities may harm the people using the medicine or eating the food.

$$\text{percentage purity} = \frac{\text{mass of the pure product}}{\text{mass of the impure product obtained}} \times 100\%$$

Sodium hydrogencarbonate, $NaHCO_3$, is used in the manufacture of some toothpastes and as a raising agent in food production. The purity of this substance can be obtained by measuring how much carbon dioxide is given off.

❓ Worked example

84 g of sodium hydrogencarbonate was thermally decomposed and 11.5 dm^3 of carbon dioxide gas was collected at room temperature and pressure (r.t.p.).

The equation for the reaction is:

$$\underset{\text{2 moles}}{2NaHCO_3(s)} \rightarrow Na_2CO_3(s) + H_2O(g) + \underset{\text{1 mole}}{CO_2(g)}$$

Step 1: Calculate the relative formula mass of sodium hydrogencarbonate (A_r: Na = 23; C = 12; O = 16; H = 1)

The relative formula mass of $NaHCO_3$ = 84

Step 2: 2 moles of $NaHCO_3$ produces 1 mole of CO_2.

168 g of $NaHCO_3$ would give 44 g of CO_2, which would have a volume of 24 dm^3 at r.t.p.

84 g of $NaHCO_3$ should give 12 dm^3 of CO_2 at r.t.p.

Hence the mass of $NaHCO_3$ in the sample was

$$84 \times \frac{11.5}{12} = 80.5 \text{ g}$$

Step 3: Calculate the percentage purity.

There is 80.5 g of sodium hydrogencarbonate in the 84 g sample.

$$\text{percentage purity} = \frac{80.5}{84} \times 100\% = 95.8\%$$

In a chemical reaction, not all the reactants are always completely used up. Usually one of the reactants will remain at the end of the reaction because it has nothing to react with. This is the excess reactant. The reactant which has been completely used up and does not remain at the end of the reaction is called the limiting reactant. The amount of product that can be obtained is determined by the limiting reactant. This is often the case when acids are used to make salts by their reactions with bases or carbonates (see Chapter 8, p. 122). An excess of the base or carbonate is used to ensure the acid has completely reacted. The acid would be the limiting reactant as it determines the amount of the salt produced.

In a chemical reaction, it is important to be able to determine which reactant is the limiting reactant as this will allow you to find the maximum mass of products which can be produced.

? Worked example

How much copper(II) oxide could be formed if 16 g of copper reacted with 20 g of oxygen gas?

Step 1: Write down the balanced chemical equation for the reaction.

$$2Cu(s) + O_2(g) \rightarrow 2CuO(s)$$

2 moles   1 mole   2 moles

Step 2: Convert the mass of each reactant into moles:

Moles of Cu used = 16/64 = 0.25 moles

Moles of O_2 used = 20/(2×16) = 0.625 moles

Step 3: Now look at the ratio of the moles in the balanced equation:

2 moles of Cu would react with 1 mole of O_2

0.25 moles of Cu would react with 0.25/2 = 0.125 moles of O_2

So Cu will be the limiting reactant because it would be used up, but there would be 0.625 – 0.125 = 0.5 moles of O_2 remaining at the end of the reaction. The oxygen would have been used in excess.

Step 4: Now we know which is the limiting reactant, we can use the mole ratio between it and the product to find the mass of copper(II) oxide we could produce.

0.25 moles of Cu would give 0.25 moles of CuO

Mass of CuO which could be produced = 0.25 × (64 + 16) = 20 g

▶ Test yourself

Use the following A_r values to answer the questions below: O = 16; Mg = 24; S = 32; K = 39; Cu = 63.5.

16 Calculate the mass of sulfur dioxide produced by burning 16 g of sulfur in an excess of oxygen.
17 Calculate the mass of sulfur which, when burned in excess oxygen, produces 640 g of sulfur dioxide.
18 Calculate the mass of copper required to produce 159 g of copper(II) oxide when heated in excess oxygen.
19 A rocket uses hydrogen as a fuel. Calculate the volume of hydrogen used to produce 24 dm³ of water ($H_2O(g)$).
20 Calculate the volume of 2 mol/dm³ solution of sulfuric acid required to react with 24 g of magnesium.
21 20 cm³ of 0.2 mol/dm³ solution of hydrochloric acid just neutralised 15 cm³ of potassium hydroxide solution in a titration (see Chapter 8, p. 128). What is the concentration of potassium hydroxide solution used in this neutralisation reaction?
22 Calculate the percentage composition of ammonium nitrate, NH_4NO_3.
23 What is the maximum mass of magnesium oxide that could be produced if 48 g of magnesium reacted with 38 g of oxygen gas?

Revision checklist

After studying Chapter 4 you should be able to:

✔ Define and, given data, calculate the molecular and empirical formula of a compound.

✔ Describe the terms relative atomic, molecular and formula mass.

✔ State that the mole is the unit of amount of substance and that it contains 6.02×10^{23} particles.

✔ Carry out calculations to find masses, moles, volumes and concentrations in reactions using solids gases and solutions.

✔ State that concentration has units of g/dm^3 and mol/dm^3, and convwert between them.

✔ Calculate percentage yield and purity.
✔ Calculate the percentage composition of a compound.

Exam-style questions

Use the data in the table below to answer the questions.

Element	A_r
H	1
C	12
N	14
O	16
Na	23
Mg	24
Si	28
S	32
Cl	35.5
Fe	56

1 Calculate the mass of:
  a 1 mole of:
    i chlorine molecules
    ii iron(III) oxide
  b 0.5 moles of:
    i magnesium nitrate
    ii ammonia.

2 Calculate the volume occupied, at r.t.p., by the following gases. (1 mole of any gas occupies a volume of 24 dm³ at r.t.p.)
  a 12.5 moles of sulfur dioxide gas
  b 0.15 mole of nitrogen gas

3 Calculate the number of moles of gas present in the following:
  a 36 cm³ of sulfur dioxide
  b 144 dm³ of hydrogen sulfide.

4 Use the following experimental information to determine the empirical formula of an oxide of silicon.

    Mass of crucible = 18.20 g

    Mass of crucible + silicon = 18.48 g

    Mass of crucible + oxide of silicon = 18.80 g

5 a Calculate the empirical formula of an organic liquid containing 26.67% of carbon and 2.22% of hydrogen, with the rest being oxygen.
  b The M_r of the liquid is 90. What is its molecular formula?

6 Iron is extracted from its ore, hematite, in a blast furnace. The main extraction reaction is:

$$Fe_2O_3(s) + 3CO(g) \rightarrow 2Fe(s) + 3CO_2(g)$$

  a Identify the reducing agent in this process.
  b Give the oxide of iron shown in the equation.
  c Explain why this is a redox reaction.
  d Calculate the mass of iron which will be produced from 640 tonnes of hematite.

5 Electrochemistry

FOCUS POINTS

★ What is the process of electrolysis?
★ How is electrolysis useful to us?
★ How do we make use of the Earth's aluminium?
★ Will fuel cells be used for transport in the future?

In this chapter, you will look at the process of electrolysis and the industries based around the process of electrolysis. You will look at the electrolysis of various substances in the molten state, and in aqueous solution. You will see how useful electrolysis is across manufacturing and fuel production industries, by considering the extraction of aluminium, chlor-alkali industry and the industry based around electroplating. Finally, you will learn about fuel cells and how these cells may be used in our methods of transport.

5.1 Electricity and chemistry

Take a look at the photographs in Figure 5.1. What do you think they all have in common?

They all involve electricity and the process known as **electrolysis**. Electrolysis is the breakdown of an ionic compound, either molten or in solution, by the passage of electricity through it. The substance that is decomposed is called the **electrolyte** (Figure 5.2). An electrolyte is a substance that conducts electricity when in the molten state or in solution.

The electricity is carried through the electrolyte by **ions**. In the molten state and in aqueous or water solution, the ions are free to move to the appropriate electrodes due to weakened forces of attraction between them.

»» Substances that do not conduct electricity when in the molten state or in solution are called non-electrolytes.
»» Substances that conduct electricity to a small extent in the molten state or in solution are called weak electrolytes.

The electric current enters and leaves the electrolyte through **electrodes**, which are usually made of unreactive metals, such as platinum or the non-metal carbon (graphite). These are said to be **inert electrodes** because they do not react with the products of electrolysis. The two electrodes are the **cathode**, the negative electrode which attracts **cations** (positively charged ions), and the **anode**, the positive electrode which attracts **anions** (negatively charged ions).

a This watch has a thin coating of gold over steel; the thin coating is produced by electrolysis
b This picture frame has been silver plated using an electroplating process involving electrolysis
c Aluminium is produced by electrolysis

▲ Figure 5.1

> **Key definition**
>
> **Electrolysis** is the decomposition of an ionic compound, when molten or in aqueous solution, by the passage of an electric current.
>
> Metals or hydrogen are formed at the cathode and non-metals (other than hydrogen) are formed at the anode.

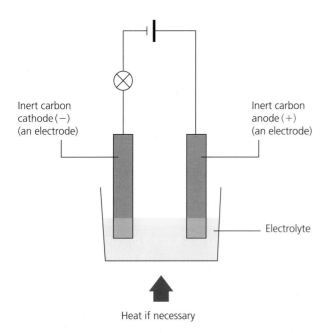

▲ **Figure 5.2** The important terms used in electrolysis

In the example of electrolysis of molten lead(II) bromide, you will see that the transfer of charge during electrolysis is by:

>> the movement of electrons in the metallic or graphite electrodes
>> the loss or gain of electrons from the external circuit at the electrodes
>> the movement of ions in the electrolyte.

The conduction that takes place in the electrodes is due to the movement of delocalised electrons (p. 47), whereas in the electrolyte the charge carriers are ions.

Electrolysis is very important within many manufacturing industries. To help you to understand how the aluminium sheet (Figure 5.1c) was obtained from its **ore**, **bauxite**, we will first consider the electrolysis of lead(II) bromide.

Key definition

The main ore of aluminium is **bauxite** and aluminium is extracted from it by electrolysis.

5.2 Electrolysis of lead(II) bromide (Teacher demonstration)

This experiment should only ever be carried out as a teacher demonstration and in a fume cupboard.

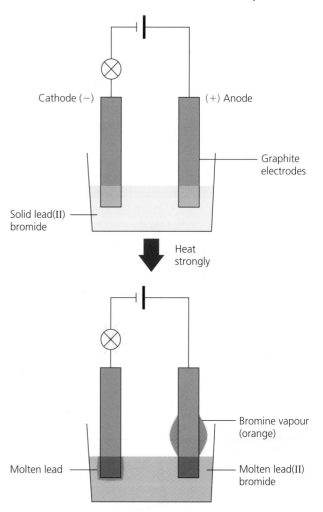

▲ **Figure 5.3** The electrolysis of molten lead(II) bromide

Consider Figure 5.3, which shows solid lead(II) bromide ($PbBr_2$) in a crucible with two carbon electrodes in contact with it. When the electrodes are first connected, the bulb does not light because the solid compound does not allow electricity to pass through it. However, when the compound is heated until it is molten, the bulb does light. The lead(II) bromide is now behaving as an electrolyte. When this happens, an orange-red vapour is seen at the anode and lead metal is produced at the cathode. Lead(II) bromide is a **binary compound**. Binary compounds are those that contain two elements chemically combined. Generally, when molten binary compounds such as lead(II) bromide are electrolysed, metals such as lead are formed at the cathode and non-metals, such as bromine, are formed at the anode.

The passage of an electric current in the process of electrolysis results in the break-up (decomposition) of lead(II) bromide into its constituent elements.

molten lead(II) bromide → bromine + lead

$$PbBr_2(l) \rightarrow Br_2(g) + Pb(l)$$

For lead metal to be formed, or deposited, at the cathode, the lead ions must be attracted to and move towards the electrode (Figure 5.4). To produce lead metal atoms, these lead ions must each collect two electrons at the cathode:

lead ion + electrons → lead atom

$$Pb^{2+}(l) + 2e^- \rightarrow Pb(l)$$

As you saw in Chapter 3 (p. 31), this process of gaining electrons is called reduction. Chemical equations such as those shown for the production of lead at the cathode are known as ionic half-equations. Similar ones can be written for the process taking place at the anode.

To form bromine molecules, each bromide ion must first of all move to the anode and lose its extra negative charge at the anode, and so form a neutral bromine atom:

Cathode (−)   Anode (+)

← Pb^{2+} Br$^-$→
Br$^-$→

Molten lead(II) bromide (PbBr$_2$)

▲ **Figure 5.4** The lead ions (Pb^{2+}) are attracted to the cathode and the bromide ions (Br$^-$) are attracted to the anode

bromide ion → bromine atom + electron

$$Br^-(l) \rightarrow Br + e^-$$

Two bromine atoms then combine to form a bromine molecule:

bromine atoms → bromine molecule

$$2Br \rightarrow Br_2(g)$$

As you saw in Chapter 3 (p. 31), this process of losing electrons is called oxidation.

? Worked example

1 a Predict the products of the electrolysis of molten magnesium fluoride (MgF$_2$).
The metal element of the magnesium fluoride is produced at the cathode. The magnesium ions (Mg^{2+}) are attracted and deposited at this electrode.
The non-metal element of the magnesium fluoride is produced at the anode. The fluoride ions (F$^-$) are attracted and deposited at this electrode.

b Write ionic half-equations for the substances produced at the different electrodes you answered in **1a**.
Ions present in MgF$_2$ are Mg^{2+} and F$^-$.
At the cathode:

$$Mg^{2+}(l) + 2e^- \rightarrow Mg(l)$$

At the anode:

$$2F^-(l) \rightarrow F_2(g) + 2e^-$$

c State whether the processes are oxidation or reduction.
The cathode reaction is reduction (Mg^{2+} gains electrons).
The anode reaction is oxidation (F$^-$ loses electrons).

▶ Test yourself

1 Copy and complete the following sentence about electrolysis using words from this list:

| compound | electricity | inert | chemical |
| platinum | molten | breakdown | |

Electrolysis is the _____ of an ionic _____ when _____ or in aqueous solution by the passage of _____ .
Electrodes are often made of graphite or _____ and are generally used in electrolysis because they are _____ .

2 a Predict the products of the electrolysis of molten:
    i   potassium chloride, KCl
    ii lead oxide, PbO.

b Write ionic half-equations for the substances produced at the different electrodes for **a i** and **a ii**.

c Use the answers to **2b** to decide which processes are oxidation and which are reduction.

5.3 Electrolysis of aluminium oxide

Aluminium is the most abundant metallic element in the Earth's crust. It was discovered in 1825 by Hans Christian Oersted in Copenhagen, Denmark, but was first isolated by Friedrich Wöhler in 1827. It makes up 8% of the crust and is found in the minerals bauxite (Figure 5.5), cryolite and mica, as well as in clay.

In the nineteenth century, Napoleon III owned a very precious dinner service. It was said to be made of a metal more precious than gold. That metal was aluminium. The reason it was precious was that it was very rarely found as the pure metal. Aluminium is a reactive metal so it is very difficult to extract from its ore. Reactive metals hold on tightly to the element(s) they have combined with and many are extracted from their ores by electrolysis.

▲ **Figure 5.5** Bauxite mining

Today we use aluminium in very large quantities. The annual production in the world is 19.5 million tonnes. The commercial extraction of aluminium has been made possible by two scientists, working independently of each other, who discovered a method using electrolysis. The two scientists were Charles Martin Hall (USA), who discovered the process in 1886, and the French chemist Paul Héroult, who discovered the process independently in the same year. The process they developed, often called the Hall–Héroult process, involves the electrolysis of aluminium oxide (alumina). The process involves the following stages.

» Bauxite, an impure form of aluminium oxide, is first treated with sodium hydroxide to obtain pure aluminium oxide, removing impurities such as iron(III) oxide and sand. This improves the conductivity of the molten aluminium oxide.

> » The purified aluminium oxide is then dissolved in molten cryolite (Na_3AlF_6). Cryolite, a mineral found naturally in Greenland, is used to reduce the working temperature of the **Hall–Héroult cell** from 2017°C (the melting point of pure aluminium oxide) to between 800 and 1000°C. Therefore, the cryolite provides a considerable saving in the energy requirements of the process.

In recent years it has become necessary to manufacture the cryolite.

» The molten mixture is then electrolysed in a cell similar to that shown in Figure 5.6.

▲ **Figure 5.6** The Hall–Héroult cell is used in industry to extract aluminium by electrolysis

The anodes of this process are blocks of carbon (graphite) which are lowered into the molten mixture from above. The cathode is the graphite lining of the steel vessel containing the cell.

Aluminium oxide is an ionic compound. When it is melted, the ions become mobile, as the strong electrostatic forces of attraction between them are broken by the input of heat energy. During electrolysis, the negatively charged oxide ions are attracted to the anode (the positive electrode), where they lose electrons (oxidation).

oxide ions → oxygen molecules + electrons

$$2O^{2-}(l) \rightarrow O_2(g) + 4e^-$$

The positive aluminium ions are attracted to the cathode (the negative electrode). They gain electrons to form molten aluminium metal (reduction).

aluminium ions → electrons + aluminium metal

$$Al^{3+}(l) \rightarrow 3e^- + Al(l)$$

A handy way of remembering what happens is **OIL RIG** – **O**xidation **I**s **L**oss, **R**eduction **I**s **G**ain of electrons.

The overall reaction which takes place in the cell is:

aluminium oxide $\xrightarrow{\text{electrolysis}}$ aluminium + oxygen

$$2Al_2O_3(l) \rightarrow 4Al(l) + 3O_2(g)$$

The molten aluminium collects at the bottom of the cell and it is siphoned out at regular intervals. No problems arise with other metals being deposited since the cryolite is largely 'unaffected' by the flow of electricity. Problems do arise, however, with the graphite anodes. At the working temperature of the cell, the oxygen liberated reacts with the graphite anodes, producing carbon dioxide.

carbon + oxygen → carbon dioxide

$$C(s) + O_2(g) \rightarrow CO_2(g)$$

The anodes burn away and have to be replaced on a regular basis.

The electrolysis of aluminium oxide is a continuous process in which vast amounts of electricity are used. Approximately 15 kWh of electricity are used to produce 1 kg of aluminium. In order to make the process an economic one, a cheap form of electricity is required. Hydroelectric power (HEP) is usually used for this process. The plant shown in Figure 5.7 uses an HEP scheme to provide some of the electrical energy required for this process.

▲ **Figure 5.7** An aluminium smelting plant

Using cheap electrical energy has allowed aluminium to be produced in such large quantities that it is the second-most widely used metal after iron. It is used in the manufacture of electrical/transmission cables. Aluminium strands are wrapped around a core of steel wires. Aluminium is used because of its low density, chemical inertness and good **electrical conductivity**. This means that it will take a high voltage and low current to carry the same power. Owing to the first two of these properties of aluminium, it is also used for making cars, bikes, cooking foil and food containers, as well as in alloys (Chapter 10, p. 165) such as duralumin, which is used in the manufacture of aeroplane bodies (Figure 5.8). Worldwide production of aluminium now exceeds 40 million tonnes each year.

▲ **Figure 5.8** Aluminium is used in the manufacture of aeroplane bodies

Recycling aluminium

The metal aluminium is the third-most abundant element in the Earth's crust after the non-metals oxygen and silicon, and the most abundant metal in the Earth's crust. However, it is more expensive to produce than many metals, such as iron, due to the cost of the large amounts of electricity which are needed for the **electrolysis** process. As a result, globally up to 50% of the aluminium used is recycled.

Obtaining aluminium from recycled scrap requires approximately 5% of the energy needed to extract aluminium from its ore. However, it should be born in mind that collecting scrap metal and transport to the **metal extraction** plant add extra costs to the overall recycling costs.

There are serious environmental problems associated with the location of aluminium plants including:

»» the effects of the extracted impurities, which form a red mud (Figure 5.9)
»» the fine cryolite dust, which is emitted through very tall chimneys so as not to affect the surrounding area
»» the claimed link between environmental aluminium and a degenerative brain disorder called Alzheimer's disease – it is thought that aluminium is a major influence on the early onset of this disease. However, the evidence is still inconclusive.

▲ **Figure 5.9** Bauxite pollution of the Amazon in Brazil

Test yourself

3 Produce a flow chart to summarise the processes involved in the extraction of aluminium metal from bauxite.
4 Explain why the mixture of gases formed at the anode contains oxygen, carbon dioxide and traces of fluorine.

Going further

Anodising

Anodising is an electrolytic process in which the surface coating of oxide on aluminium (Al_2O_3) is made thicker. In this process, the aluminium object is made the anode in a cell in which the electrolyte is dilute sulfuric acid. During the electrolysis process, oxygen is produced at the anode and combines with the aluminium. The oxide layer on the surface of the aluminium therefore increases. Dyes can be mixed with the electrolyte and so the new thicker coating of oxide is colourful and also decorative (Figure 5.10).

▲ **Figure 5.10** The oxide layer on the surface of these aluminium cups has been thickened, and dyes added to obtain the vibrant colours

Test yourself

5 Complete the following passage about the extraction of aluminium using words from the list below.

anode   cell   cryolite   replaced   burn
cathode   oxygen   bauxite   carbon

Aluminium is extracted from its molten ore, _____ by electrolysis in an electrolytic _____ . Before the extraction process begins, _____ is added. The _____ electrodes are then lowered into the cell. During the process aluminium is produced at the _____ and is siphoned off. At the _____ , oxygen is produced. Because of the high temperature required in the process, the carbon electrodes _____ away due to the reaction of the anodes with _____ produced at that electrode. Therefore the anodes have to be _____ regularly.

5.4 Electrolysis of aqueous solutions

Other industrial processes involve the electrolysis of aqueous solutions. To explain what is happening in these processes, we will first consider the electrolysis of dilute sulfuric acid.

Electrolysis of dilute sulfuric acid

Pure water is a very poor conductor of electricity because there are so few ions in it. However, it can be made to decompose if an electric current is passed through it in a Hofmann voltameter, as in Figure 5.11.

Oxygen collected here — Hydrogen collected here

Water (with a little dilute sulfuric acid added to increase its conductivity)

Platinum electrodes (inert)

Anode (+)   Cathode (−)

Power supply

▲ **Figure 5.11** A Hofmann voltameter used to electrolyse water

To enable water to conduct electricity better, some dilute sulfuric acid (or sodium hydroxide solution) is added. When the power is turned on and an electric current flows through this solution, gases can be seen to be produced at the two electrodes and they are collected in the side arms of the apparatus. After about 20 minutes, roughly twice as much gas is produced at the cathode as at the anode.

The gas collected at the cathode burns with a squeaky pop, showing it to be hydrogen gas (Table 14.6, p. 236).

> For hydrogen to be collected in this way, the positively charged hydrogen ions must have moved to the cathode.
>
> hydrogen ions + electrons → hydrogen molecules
>
> $$4H^+(aq) + 4e^- \rightarrow 2H_2(g)$$
>
> If, during this process, the water molecules lose $H^+(aq)$, then the remaining portion must be hydroxide ions, $OH^-(aq)$. These ions are attracted to the anode.

The gas collected at the anode relights a glowing splint, showing it to be oxygen (Table 14.6, p. 236).

> This gas is produced in the following way.
>
> hydroxide ions → water + oxygen + electrons
>
> $$4OH^-(aq) \rightarrow 2H_2O(l) + O_2(g) + 4e^-$$

Generally when aqueous solutions are electrolysed, hydrogen is produced at the cathode and non-metals (other than hydrogen) are formed at the anode.

This experiment was first carried out by Sir Humphry Davy. It confirmed that the formula for water was H_2O.

In the electrolysis of dilute sulfuric acid, platinum (an inert electrode) may be replaced by carbon (graphite). The only difference to occur is that, as well as oxygen being produced at the anode, a little carbon dioxide will also be formed.

Electrolysis of concentrated aqueous sodium chloride

When sodium chloride (NaCl) is dissolved in water, the ions present become free to move. When concentrated aqueous sodium chloride is electrolysed, the two gases, hydrogen and chlorine, are produced. Sodium might have been expected to be produced at the cathode. Why does this not happen?

Four ions are present in the solution:

» from water: H^+ and OH^-
» from sodium chloride: Na^+ and Cl^-.

When the current flows, the H^+ ions and the Na^+ ions are both attracted to the cathode. The H^+ ions, however, accept electrons more easily than the Na^+ ions and so hydrogen gas (H_2) is produced at the cathode.

hydrogen ions + electrons $\xrightarrow{\text{reduction}}$ hydrogen molecules

$$2H^+(aq) + 2e^- \rightarrow H_2(g)$$

Because the sodium is more reactive than the hydrogen, it loses electrons more easily. Generally it is found that, during the electrolysis of aqueous solutions, very reactive metals like sodium are not formed at the cathode. You will see in Chapter 10 that reactive metals liberate hydrogen when they react with water.

OH^- and Cl^- are attracted to the anode. The Cl^- ions give up electrons more readily than the OH^- ions and so chlorine gas is produced by the electrode process.

chloride ions $\xrightarrow{\text{oxidation}}$ chlorine molecules + electrons

$$2Cl^-(aq) \rightarrow Cl_2(g) + 2e^-$$

This leaves a high concentration of hydroxide ions, $OH^-(aq)$, around the cathode. The solution in the region of the anode therefore becomes alkaline.

Electrolysis guidelines

The following points may help you predict the products of electrolysis in unfamiliar situations. They will also help you remember what happens at each electrode.

» Non-metals are produced at the anode whereas metals and hydrogen gas are produced at the cathode.
» At an inert anode, chlorine, bromine and iodine (the halogens) are produced in preference to oxygen.
» At an inert cathode, hydrogen is produced in preference to metals unless unreactive metals such as copper are present.

Going further

The chlor-alkali industry

The electrolysis of concentrated or saturated sodium chloride solution (brine) is the basis of a major industry. In countries where rock salt (sodium chloride) is found underground, it is mined. In other countries it can be obtained by evaporation of sea water in large, shallow lakes. Three important substances are produced in this electrolysis process – chlorine, sodium hydroxide and hydrogen. The electrolytic process is very expensive, requiring vast amounts of electricity, and the process is economical only because all three products (hydrogen, chlorine and sodium hydroxide) have a large number of uses (Figure 5.12).

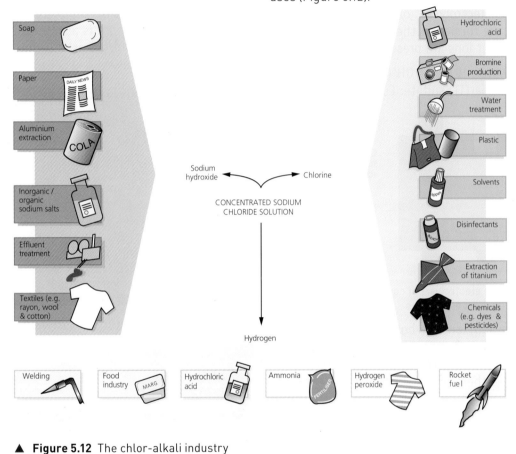

▲ **Figure 5.12** The chlor-alkali industry

❓ Worked example

1 Predict the products at the cathode and anode of the electrolysis of:
   a dilute aqueous potassium chloride
      At cathode: hydrogen (from the water)
      At anode: chlorine and oxygen (from the water).
   b concentrated hydrobromic acid (HBr)
      Concentrated acid contains very little or no water.
      So in the electrolysis of the acid, the ions present are H^+ and Br^-. Hence:
      At cathode: hydrogen
      At anode: bromine.

▶ Test yourself

6 Suggest a reason for only 'roughly' twice as much hydrogen gas being produced at the cathode as oxygen gas at the anode in the electrolysis of water.
7 The uses of sodium hydroxide can be separated on a percentage basis as follows:
   Neutralisation: 5%
   Paper manufacture: 5%
   Oil refining: 5%
   Soap/detergents: 5%
   Manufacture of rayon and acetate fibres: 16%
   Manufacture of chemicals: 30%
   Miscellaneous uses: 34%
   Use a graph-plotting program to create a pie chart of these data.

5.5 Electrolysis of copper(II) sulfate aqueous solution

Copper(II) sulfate solution ($CuSO_4(aq)$) may be electrolysed using inert graphite electrodes in a cell similar to that shown in Figure 5.13. When the solution is electrolysed, oxygen gas and copper metal are formed at the anode and cathode respectively. Four ions are present in solution:

» from the water: $H^+(aq) + OH^-(aq)$
» from the copper(II) sulfate: $Cu^{2+}(aq) + SO_4^{2-}(aq)$.

Oxygen

Copper(II) sulfate solution

Copper forms on surface of carbon electrode

Bung

+   −

▲ **Figure 5.13** The electrolysis of copper(II) sulfate solution using inert electrodes

$H^+(aq)$ and $Cu^{2+}(aq)$ ions are both attracted to the cathode, the Cu^{2+} ions accepting electrons more readily than the H^+ ions (preferential discharge). Copper metal is therefore deposited at the cathode (Figure 5.13).

copper ions + electrons → copper atoms

$$Cu^{2+}(aq) + 2e^- \rightarrow Cu(s)$$

Anode   Cathode

e^- OH

e^- OH

Electrons from OH^- left behind e^- OH   → O_2 +$2H_2O$

e^- OH

$Cu \leftarrow Cu^{2+}$ +$2e^-$ Electrons collected by Cu^{2+} ions. Copper coats the electrode

$Cu \leftarrow Cu^{2+}$ +$2e^-$

▲ **Figure 5.14** Oxygen is given off at the anode and copper is deposited at the cathode

$OH^-(aq)$ and $SO_4^{2-}(aq)$ ions are both attracted to the anode. The OH^- ions release electrons more easily than the SO_4^{2-} ions, so oxygen gas and water are produced at the anode (Figure 5.14).

hydroxide ions → oxygen + water + electrons

$$4OH^-(aq) \rightarrow O_2(g) + 2H_2O(l) + 4e^-$$

Purification of copper

As copper is a very good conductor of electricity, it is used for electrical wiring and cables (Figure 5.15). Pure copper is also used in the manufacture of cooking utensils owing to its high thermal conductivity.

a The copper used in electrical wiring has to be very pure

b Due to the high density of copper and its cost, steel-cored aluminium cables are used for electrical transmission through national grids

▲ **Figure 5.15**

However, even small amounts of impurities cut down this conductivity quite noticeably, whether in fine wires or larger cables. The metal must be 99.99% pure to be used in this way. To ensure this level of purity, newly extracted copper has to be purified by electrolysis.

The impure copper is used as the anode in the electrolysis process, and is typically 1 m square, 35–50 mm thick and 330 kg in weight. The cathode is a 1 mm thick sheet and weighs about 5 kg; it is made from very pure copper. Because copper is itself involved in the electrolytic process, the copper cathode is known as an 'active' electrode. The electrolyte is a solution of copper(II) sulfate (0.3 mol/dm³) acidified with a 2 mol/dm³ solution of sulfuric acid to help the solution conduct electricity (Figure 5.16).

Cathode (−)   Anode (+)

Pure copper

Impure copper

Acidified copper(II) sulfate solution (electrolyte)

(−)   (+)

Impurities, such as the precious metals gold and silver, appear in the anode slime

▲ **Figure 5.16** Copper purification process

When the current flows, the copper moves from the impure anode to the pure cathode. Any impurities fall to the bottom of the cell and collect below the anode in the form of a slime. This slime is rich in precious metals and the recovery of these metals is an important aspect of the economics of the process. The electrolysis proceeds for about three weeks until the anodes are reduced to about 10% of their original size and the cathodes weigh between 100 and 120 kg. A potential of 0.25 V and a current density of 200 A/m² are usually used.

The ions present in the solution are:

» from the water: $H^+(aq) + OH^-(aq)$
» from the copper(II) sulfate: $Cu^{2+}(aq) + SO_4^{2-}(aq)$.

During the process, the impure anode loses mass because the copper atoms lose electrons and become copper ions, $Cu^{2+}(aq)$ (Figure 5.17).

copper atoms → copper ions + electrons

$$Cu(s) \rightarrow Cu^{2+}(aq) + 2e^-$$

The electrons released at the anode travel around the external circuit to the cathode. There the electrons are passed on to the copper ions, $Cu^{2+}(aq)$, from the copper(II) sulfate solution and the copper is deposited or copper plated on to the cathode.

copper ions + electrons → copper atoms

$$Cu^{2+}(aq) + 2e^- \rightarrow Cu(s)$$

Cathode (−)   Anode (+)

Pure copper electrode

Impure copper electrode

Electrolyte of acidified copper(II) sulfate solution

▲ **Figure 5.17** The movement of ions in the purification of copper by electrolysis

Going further

The annual production of copper worldwide is in excess of 16 million tonnes. However, a large amount of the copper we need is obtained by recycling. This way of obtaining copper is increasing in importance as it becomes more difficult and expensive to locate and extract the copper ore from the ground.

 Practical skills

Electrolysis of copper sulfate solution

For safe experiments/demonstrations which are related to this chapter, please refer to the *Cambridge IGCSE Chemistry Practical Skills Workbook*, which is also part of this series.

Safety

- Eye protection must be worn.
- Take care with copper(II) sulfate solution: ≤ 1.0 mol/dm^3 is an irritant and corrosive.

As part of a project about purifying metals, a student wanted to investigate the purification of copper. Their teacher gave them a piece of impure copper and a piece of pure copper foil, as well as some sulfuric acid acidified copper(II) sulfate solution. The apparatus available is shown in Figure 5.16. The electrical energy was provided by a power supply. The student then followed these directions:

1 Weigh the copper pieces separately and enter the masses in your table of results. You should also record your observations after the process is complete.
2 Set up the apparatus as shown in Figure 5.16, adding 50 cm^3 of the acidified copper(II) sulfate to the beaker.
3 Switch on the power supply and let the process run for 15 minutes.
4 Switch off the power supply, carefully dry the copper strips and reweigh the copper strips.

This is the table of results obtained by the student.

Copper strip	Mass at start/g	Mass after 15 minutes/g	Change in mass/g	Observations
Pure copper cathode(-)	3.16	3.40		Has an orange/brown colouration on the part of the copper that has been in the copper(II) sulfate solution
Impure copper anode (+)	3.25	2.98		Looks a little thinner. Small pile of material below this electrode

a What safety precautions would you take in carrying out the experiment?
b Why do you think the copper(II) sulfate was acidified?
c The colour of the copper(II) sulfate solution does not change. Why not?
d Calculate the mass change for each of the copper electrodes.
e Are the changes in mass what you expected? Explain your answer.
f It was found that there was some material below the anode. Explain how you could obtain a sample of this material.
g It was thought that this mixture you have isolated in **f** was a mixture of substances including what appeared to be a dye.
  i Which technique could be used to check this?
  ii Design an experiment, using the technique you have described in **i**, which would allow you to identify the components of the dye material.

5.6 Fuel cells

Scientists have found a way of changing chemical energy into electrical energy, using a **fuel cell** (Figure 5.18). Fuel cells are chemical cells except that the reagents are supplied continuously to the electrodes. The reagents are usually hydrogen and oxygen. The fuel cell principle was first discovered by Sir William Grove in 1839.

▲ **Figure 5.18** The space shuttle's computers use electricity produced by fuel cells

When Grove was electrolysing water and he switched off the power supply, he noticed that a current still flowed but in the reverse direction. Subsequently, the process was explained in terms of the reactions at the electrodes' surfaces of the oxygen and hydrogen gases which had been produced during the electrolysis.

The hydrogen fuel cells used by NASA in the US space programme are about 70% efficient and, since the only product is water, they are pollution free.

> **Key definition**
>
> A hydrogen–oxygen **fuel cell** uses hydrogen and oxygen to produce electricity with water as the only substance produced.

The aqueous NaOH electrolyte is kept within the cell by electrodes which are porous, allowing the transfer of O_2, H_2 and water through them (Figure 5.19). As O_2 gas is passed into the cathode region of the cell, it is reduced:

$$O_2(g) + 2H_2O(l) + 4e^- \rightarrow 4OH^-(aq)$$

The OH^- ions formed are removed from the fuel cell by reaction with H_2:

$$H_2(g) + 2OH^-(aq) \rightarrow 2H_2O(l) + 2e^-$$

The electrons produced by this process pass around an external circuit to the cathode.

▲ **Figure 5.19** A diagrammatic view of a fuel cell

Advantages and disadvantages of the hydrogen fuel cell

As issues surrounding climate change gain momentum, people and governments all around the world are growing more and more concerned each day. Countries all over the world are considering cleaner and more sustainable sources of power. There is adoption of renewable technologies, such as electric vehicles. Another technology that is gaining interest worldwide is fuel cell technology. Fuel cell technology has its advantages; for example, a fuel cell:

» uses hydrogen and oxygen and makes non-polluting water in the process of generating electricity, whereas petrol and diesel engines produce many pollutants
» is similar to a battery but does not require any external charging
» is capable of producing electricity as long as hydrogen fuel and oxygen are supplied.

However there are some disadvantages to using fuel cells. For example:

» hydrogen is in the gas state at room temperature and pressure, so it is difficult to store in a car
» the infrastructure does not yet exist, as it does for fossil fuels, for example, the number of refuelling stations
» fuel cells and electric motors are less durable than petrol or diesel engines, which means they do not last as long
» fuel cells are very expensive at the moment.

▶ Test yourself

12 The fuel cell was discovered during electrolysis experiments with water. It is the reverse process which produces the electricity. Write a balanced chemical equation to represent the overall reaction taking place in a fuel cell.

5.7 Electroplating

Electroplating is the process involving electrolysis where one metal is plated, or coated, with another. Often the purpose of **electroplating** is to improve the appearance of the object and give a protective coating to the metal beneath. For example, bath taps are chromium plated to prevent corrosion, and at the same time are given a shiny, more attractive finish (Figure 5.20).

> **Key definition**
> **Electroplating** is applied to metals to improve their appearance and resistance to corrosion.

▲ **Figure 5.20** This tap has been chromium plated

The electroplating process is carried out in a cell such as the one shown in Figure 5.21a. This is often known as the 'plating bath' and it contains a suitable electrolyte, usually a solution of a metal salt.

For silver plating, the electrolyte is a solution of a silver salt. The item to be plated is made the cathode in the cell so that the metal ions move to it when the current is switched on. The cathode reaction in this process is:

silver ions + electrons → silver atoms

$$Ag^+(aq) \quad + \quad e- \quad \rightarrow \quad Ag(s)$$

Anode (+)   Cathode (−)

Object to be plated,
e.g. spoon

Anode made
from the metal
being used
for plating,
e.g. silver

Electrolyte
containing the
metal being used
for plating,
e.g. silver

a   Silver plating a spoon

The Ag⁺ ions are attracted to the
cathode, where they gain electrons

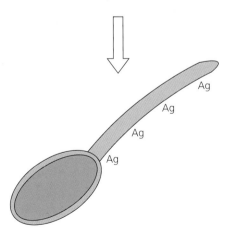

A coating of silver forms
on the spoon at the cathode

b   Explaining silver plating

▲ **Figure 5.21**

→ **Going further**

Plating plastics

Nowadays it is not only metals that are electroplated. Plastics have been developed that are able to conduct electricity. For example, the plastic poly(pyrrole) can be electroplated in the same way as the metals we have discussed earlier (Figure 5.22).

▲ **Figure 5.22** This plastic has been coated with copper by electrolysis

▲ **Figure 5.23** This leaf has been electroplated

Test yourself

13 Explain why copper(II) chloride solution would not be used as an electrolyte in the electrolyte cell used for copper plating.

14 Write equations which represent the discharge at the cathode of the following ions:
a K^+ **b** Pb^{2+} **c** Al^{3+}
and at the anode of:
d Br^- **e** O^{2-} **f** F^-.

Revision checklist

After studying Chapter 5, you should be able to:
✓ Define electrolysis.
✓ In a simple electrolytic cell, identify the component parts of anode, cathode and the electrolyte.
✓ Identify which electrode in a simple electrolytic cell the anions and cations are attracted to.

✓ Describe the transfer of charge that takes place during electrolysis.
✓ Construct ionic half-equations for reactions at the anode (to show oxidation) and at the cathode (to show reduction).

✓ Describe the electrode products as well as the observations made during the electrolysis of the following substances: molten lead(II) bromide, concentrated aqueous sodium chloride, dilute sulfuric acid.

✓ Describe the process of the extraction of aluminium from purified bauxite.

✓ State that metals or hydrogen are formed at the cathode and that non-metals (other than hydrogen) are formed at the anode.
✓ Predict the identity of the products of electrolysis of a specified binary compound, such as lead(II) bromide, in the molten state.

✓ Describe how the refining of copper takes place by the process of electrolysis using an aqueous solution of copper(II) sulfate with carbon and copper electrodes.

✓ Describe how and why metals are electroplated.
✓ State that hydrogen–oxygen fuel cells use hydrogen and oxygen to generate electricity with water as the only substance produced.

✓ Describe the advantages and disadvantages of using hydrogen–oxygen fuel cells in comparison with gasoline or petrol engines in vehicles.

Exam-style questions

1

Dilute sulfuric acid

This is a diagram of an experiment in which electricity was passed through a mixture of distilled water containing a little dilute sulfuric acid.

a Identify the gas that collects at **A**. [1]
b Identify the gas that collects at **B**. [1]
c If 100 cm³ of gas collects in **A**, how much would there be in **B**? [1]
d Identify the metal usually used for **X** and **Y**. [1]
e **X** is called the _____ . [1]
f **Y** is called the _____ . [1]
g Give the formulae of the three ions present in the solution. [3]

h Give the ionic half-equations for the reactions that take place at both X and Y. [4]

2 Explain the meaning of each of the following terms.
a Anode [2]
b Cathode [2]
c Electrolysis [2]
d Electrolyte [2]
e Anion [2]
f Cation [2]

g Oxidation, in terms of electrons [1]
h Reduction, in terms of electrons [1]

3 Copper is purified by electrolysis, as in the example shown below.

a Identify the materials used for the electrodes **A** and **B**. [2]
b Identify the electrolyte **C** and substance **D**. [2]
c Explain why substance **D** is of economic importance in respect of this process. [2]
d Give word and symbol equations for the reactions which take place at the cathode and anode during this process. [6]
e Explain why electrolyte **C** has to be acidified with dilute sulfuric acid. [2]
f Explain why copper has to be 99.99% pure for use in electrical cables. [2]

4 Copy and complete the table below, which shows the results of the electrolysis of four substances using inert electrodes.

Electrolyte	Product at anode (positive electrode)	Product at cathode (negative electrode)
Molten aluminium oxide		Aluminium
Concentrated sodium chloride solution	Chlorine	
Molten lithium chloride		
Silver nitrate solution		Silver

[5]

a Give the meaning of 'inert electrodes'. [2]
b Explain why the sodium chloride solution becomes progressively more alkaline during electrolysis. [3]
c Explain why solid lithium chloride is a non-conductor of electricity, whereas molten lithium chloride and lithium chloride solution are good conductors of electricity. [3]

d During the electrolysis of molten aluminium chloride ($AlCl_3$) the carbon anodes are burned away. Explain why this happens and write balanced chemical equations for the reactions that take place. [4]

5 Sodium hydroxide is made by the electrolysis of brine (concentrated sodium chloride solution).

> a Give ionic half-equations for the reactions which take place at the cathode and anode. State clearly whether a reaction is oxidation or reduction. [8]

b Give two large-scale uses of the products of this electrolytic process. [6]

c Analyse the following statement: 'This electrolytic process is a very expensive one.' [2]

6 Electroplating is an important industrial process.
a Explain what electroplating is. [2]
b Explain why certain metals are electroplated. [2]
c Give two examples of the use of electroplating. [2]

Chemical energetics

In this chapter, you will look at the energy, or enthalpy, changes which happen when a chemical reaction occurs. We know that when we burn a fuel, energy is released, and the surroundings get warmer. Many of the fuels we use are obtained from the fossil fuel petroleum. How are important fuels such as gasoline or diesel obtained from petroleum?

In this chapter, we will look at why heat is produced and you will be introduced to the idea of chemical reactions which occur and cause things to get cooler. Have you met any of these reactions? What could they be used for?

By the end of this chapter, you should be able to calculate the energy change which occurs when a chemical reaction happens.

6.1 Substances from petroleum

What do the modes of transport in Figure 6.1 have in common?

▲ **Figure 6.1** What type of fuel do these modes of transport use?

They all use liquids obtained from **petroleum** as fuels.

Key definition
Petroleum is a mixture of hydrocarbons.

Oil refining

Petroleum (also called crude oil) is a complex mixture of compounds known as **hydrocarbons** (Figure 6.2). Hydrocarbons are molecules which contain only the elements hydrogen and carbon bonded together covalently (Chapter 3, p. 39). These carbon compounds form the basis of a group called **organic compounds**. All living things are made from organic compounds based on chains of carbon atoms similar to those found in petroleum. Petroleum is not only a major source of fuel but also a raw material of enormous importance. It supplies a large and diverse chemical industry to make dozens of products that we use in everyday life, such as polymers and medicines.

> **Key definition**
>
> **Hydrocarbons** are compounds that contain hydrogen and carbon only.

▲ **Figure 6.2** Petroleum is a mixture of hydrocarbons

Petroleum is not very useful to us until it has been processed. The process, known as **oil refining**, is carried out at an oil refinery (Figure 6.3).

Refining involves separating petroleum into various batches or fractions. Chemists use a technique called **fractional distillation** to separate the different fractions. This process works in a similar way to that discussed in Chapter 14, p. 228, for separating ethanol (alcohol) and water. The different components (fractions) separate because they have different boiling points. The petroleum is heated to about 400°C to vaporise all the different parts of

▲ **Figure 6.3** An oil refinery

the mixture. The mixture of vapours passes into the fractionating column near the bottom (Figure 6.4). Each fraction is obtained by collecting hydrocarbon molecules, which have a boiling point in a given range of temperatures. For example, the fraction we know as petrol contains molecules which have boiling points between 30°C and 110°C. The molecules in this fraction contain between five and ten carbon atoms. These smaller molecules with the shorter chain lengths have lower boiling points and condense higher up the tower. Bigger hydrocarbon molecules, with longer chain lengths, have higher boiling points and condense in the lower half of the tower.

The liquids condensing at different levels are collected on trays. In this way the petroleum is separated into different fractions. This process is known as fractionation. These fractions usually contain a number of different hydrocarbons. The individual single hydrocarbons can then be obtained, again by refining the fraction by further distillation.

It is important to realise that the uses of the fractions depend on their properties. For example, one of the lower fractions, which boils in the range 250–350°C, is quite thick and sticky, and makes a good lubricant, whereas the gasoline or petrol fraction burns very easily, which makes it a good fuel for use in engines. Figure 6.4b shows the uses of the fractions.

In general, the properties of the fractions obtained from this fractional distillation change from the bottom to the top of the tower with:

» lowering boiling points
» higher volatility
» lower viscosity
» decreasing chain length.

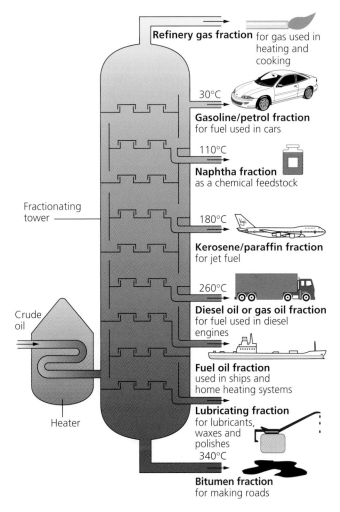

a Fractional distillation of petroleum in a refinery

▲ **Figure 6.4**

b Uses of the different fractions obtained from petroleum

▶ Test yourself

1 What do you understand by the term 'hydrocarbon'?
2 All living organisms are composed of compounds which contain carbon. Why do you think carbon chemistry is often called 'organic chemistry'?
3 List the main fractions obtained by separating the petroleum mixture and explain how they are obtained in a refinery.
4 Which of the fractions obtained from petroleum could be used for producing:

a waxes
b fuel for aeroplanes
c polymers?

5 Describe how the properties of fractions obtained from petroleum change as they are collected in a fractionating column from the bottom to the top of the column, in terms of their:

a chain length
b volatility
c viscosity
d boiling points.

6.2 Fossil fuels

Coal, petroleum and natural gas are all examples of **fossil fuels**. The term 'fossil fuels' is derived from the fact that they are formed from dead plants and animals which were fossilised over 200 million years ago during the carboniferous era (Figure 6.5).

 Going further

Coal was produced by the action of pressure and heat on dead wood from ancient forests. When dead trees fell into the swamps they were buried by mud. Over millions of years, due to movement of the Earth's crust as well as changes in climate, the land sank and the decaying wood became covered by even more layers of mud and sand. As time passed, the gradually forming coal became more and more compressed as other material was laid down above it (Figure 6.5).

Petroleum and gas were formed during the same period as coal. It is believed petroleum and gas were formed from the remains of plants, animals and bacteria that once lived in seas and lakes. This material sank to the bottom of these seas and lakes, and became covered in mud, sand and silt which thickened with time.

Anaerobic decay took place and, as the mud layers built up, high temperatures and pressures were created which converted the material slowly into petroleum and gas. As rock formed, Earth movements caused it to buckle and split, and the oil and gas were trapped in folds beneath layers of non-porous rock or cap-rock (Figure 6.6).

▲ **Figure 6.5** Piece of coal showing a fossilised leaf

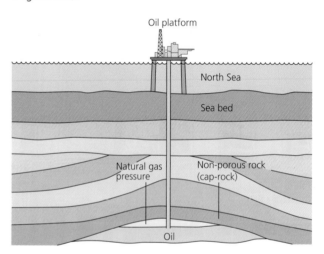

▲ **Figure 6.6** Natural gas and petroleum are trapped under non-porous rock

Test yourself

6 Coal, petroleum and natural gas are all termed 'fossil fuels'. Why is the word 'fossil' used in this context?

7 Draw a flow diagram to represent the formation of coal, oil or gas.

6.3 What is a fuel?

A fuel is any substance which can be conveniently used as a source of energy. Fossil fuels release energy in the form of heat when they undergo **combustion**.

$$\text{fossil fuel} + \text{oxygen} \rightarrow \text{carbon} + \text{water} + (\Delta H)$$
$$\text{dioxide}$$

For example, natural gas burns readily in air (Chapter 12, p. 191).

$$\text{methane} + \text{oxygen} \rightarrow \text{carbon} + \text{water} + (\Delta H)$$
$$\text{dioxide}$$

$$CH_4(g) + 2O_2(g) \rightarrow CO_2(g) + 2H_2O(l)$$

Natural gas, like petroleum, is a mixture of hydrocarbons. Methane is the main constituent of natural gas; it also contains ethane and propane, and may also contain some sulfur. The sulfur content varies from source to source; natural gas obtained from the North Sea is quite low in sulfur.

The perfect fuel would:

» be cheap
» be available in large quantities
» be safe to store and transport
» be easy to ignite and burn, causing no pollution
» release no greenhouse gases
» be capable of releasing large amounts of energy.

Solid fuels are safer than volatile liquid fuels, like petrol, and gaseous fuels, like natural gas.

Test yourself

8 'We have not yet found the perfect fuel.' Discuss this statement.

 Going further

How are fossil fuels used?

A major use of fossil fuels is in the production of electricity. Coal and natural gas are burned in power stations to heat water to produce steam, which is then used to drive large turbines (Figure 6.7). A high proportion of the electricity generated worldwide is derived from fossil fuels. Coal is becoming less important while natural gas is increasingly important.

In a power station, the turbine drives a generator to produce electricity, which is then fed into the national grid (Figure 6.7). The national grid is a system for distributing electricity throughout a country.

▲ **Figure 6.7** The way in which fuels are used to produce electricity

Other major uses of fossil fuels are:
● as major feedstock (raw material) for the chemicals and pharmaceuticals industries
● for domestic and industrial heating and cooking
● as fuels for various forms of vehicle transport.

6.4 Alternative sources to fossil fuels

Fossil fuels are an example of **non-renewable** resources, so called because they are not being replaced at the same rate as they are being used up. It is important to use non-renewable fuels very carefully and to consider alternative, **renewable** sources of energy for use now and in the future.
These include:

» nuclear
» hydroelectric
» biomass and biogas
» hydrogen
» wind

» tidal
» wave
» geothermal
» solar.

Test yourself

9 Choose two alternative sources of energy and suggest some advantages and disadvantages of each.
10 What is meant by the terms:
   a non-renewable energy sources?
   b renewable energy sources?

6.5 Exothermic and endothermic reactions

We obtain our energy needs from the combustion of fuels.

Combustion

When natural gas burns in a plentiful supply of air, it produces a large amount of energy.

methane + oxygen → carbon dioxide + water + (ΔH)

$$CH_4(g) + 2O_2(g) \rightarrow \quad CO_2(g) \quad + 2H_2O(l) + (\Delta H)$$

During this process, the complete combustion of methane, heat is given out. It is an **exothermic** reaction. An exothermic reaction is one which transfers thermal energy to the surroundings when the reaction occurs. If only a limited supply of air is available then the reaction is not as exothermic and the poisonous gas carbon monoxide is produced.

$$\text{methane} + \text{oxygen} \rightarrow \text{carbon monoxide} + \text{water} + (\Delta H)$$

$$2CH_4(g) + 3O_2(g) \rightarrow 2CO(g) + 4H_2O(l) + (\Delta H)$$

This is the incomplete combustion of methane.

The energy changes that take place during a chemical reaction can be shown by a reaction pathway diagram. Figure 6.8 shows the reaction pathway diagram for the exothermic reaction of the complete combustion of methane. The majority of chemical reactions are exothermic reactions: they give out thermal energy when they occur.

▲ **Figure 6.8** A reaction pathway diagram for the complete combustion of methane

You may have used a hand warmer to keep your hands warm in cold weather. The hand warmer contains chemicals which, when mixed together, cause an exothermic reaction to occur, giving out heat.

There are also reactions which take in thermal energy and these are known as **endothermic** reactions. When an endothermic reaction occurs, energy is taken in from the surroundings and the temperature of the surroundings decreases. An example of a use of an endothermic reaction is when a cold pack is used to treat a sprain. The cold pack contains an inner bag of water and a chemical which will react with it to produce an endothermic reaction. To use the cold pack, you squeeze the pack, which allows the water to react with the chemical and, when it is applied to the sprained area of the body, the temperature decreases.

When any reaction occurs, the chemical bonds in the reactants have to be broken – this requires energy. This is an endothermic process. When the new bonds in the products are formed, energy is given out (Figure 6.9). This is an exothermic process. The **bond energy** is defined as the amount of energy in kilojoules (kJ) associated with the breaking or making of 1 mole of chemical bonds in a molecular element or compound.

▲ **Figure 6.9** Breaking and forming bonds during the combustion of methane

Using the data in Table 6.1, which tells us how much energy is needed to break a chemical bond and how much is given out when it forms, we can calculate how much energy is involved in each stage.

▼ **Table 6.1** Bond energy data

Bond	Bond energy/kJ/mol
C—H	435
O=O	497
C=O	803
H—O	464
C—C	347
C—O	358

Bond breaking

The first step is to calculate the amount of energy needed to break the bonds in the chemical reactants, methane and oxygen gas. In this reaction, 1 mole of methane reacts with 2 moles of oxygen gas.

Breaking 4 C–H bonds in 1 mole of methane requires:

$4 \times 435 = 1740$ kJ

Breaking 2 O=O bonds in the 2 moles of oxygen requires:

$2 \times 497 = 994$ kJ

Total = 2734 kJ of energy

This is the total amount of energy required to break the bonds in the reactants.

Making bonds

The second step is to calculate the amount of energy which will be given out when the products are made. In this reaction, 1 mole of carbon dioxide and 2 moles of water are produced.

Making 2 C=O bonds in 1 mole of carbon dioxide gives out:

$2 \times 803 = 1606$ kJ

Making 4 H–O bonds in the 2 moles of water gives out:

$4 \times 464 = 1856$ kJ

Total = 3462 kJ of energy

This is the total amount of energy given out when the bonds in the products are formed.

$$\text{energy difference} = \text{energy required to break bonds} - \text{energy given out when bonds are made}$$

$$= 2734 - 3462$$

$$= -728 \text{ kJ/mol of methane burned}$$

The negative sign shows that the chemicals are losing energy to the surroundings: it is an exothermic reaction. A positive sign would indicate that the chemicals are gaining energy from the surroundings in an endothermic reaction.

The energy stored in the bonds is called the **enthalpy** and is given the symbol H. The change in energy going from reactants to products is called the **enthalpy change of reaction** and is shown as ΔH (pronounced 'delta H').

For an exothermic reaction ΔH is negative and for an endothermic reaction ΔH is positive.

> **Key definition**
>
> The transfer of thermal energy during a reaction is called the **enthalpy change, ΔH, of the reaction**.
>
> Bond breaking is an endothermic process and bond making is an exothermic process. The enthalpy change of a reaction is the difference between the energy required to break bonds and the energy given out when bonds are made.

When fuels, such as methane, are burned they require energy to start the chemical reaction. This minimum amount of energy is known as the **activation energy**, E_a (Figure 6.10). It is the minimum amount of energy which is needed to allow the colliding particles in the reaction mixture to form products.

> **Key definition**
>
> The **activation energy**, E_a, is the minimum energy that colliding particles must have in order to react.

In the case of methane reacting with oxygen, the activation energy is some of the energy involved in the initial bond breaking, as the reaction pathway in Figure 6.10 shows. The value of the activation energy varies from fuel to fuel.

▲ **Figure 6.10** A reaction pathway diagram for methane with oxygen

▲ **Figure 6.11** A reaction pathway diagram for nitrogen with oxygen

Endothermic reactions are much less common than exothermic ones. In this type of reaction, energy is absorbed from the surroundings so that the energy of the products is greater than that of the reactants. The reaction between nitrogen and oxygen gases is endothermic (Figure 6.11), and the reaction is favoured (more likely to occur) at high temperatures.

nitrogen + oxygen → nitrogen(II) oxide

$$N_2(g) \ + \ O_2(g) \ \rightarrow 2NO(g) \ \Delta H = +181 \text{ kJ/mol}$$

Dissolving is often an endothermic process. For example, when ammonium nitrate dissolves in water, the temperature of the water falls, indicating that energy is being taken from the surroundings. Photosynthesis and thermal decomposition are other examples of endothermic processes.

In equations it is usual to express the ΔH value in units of kJ/mol. For example:

$$CH_4(g) + 2O_2(g) \rightarrow CO_2(g) + 2H_2O(l) \ \Delta H = -728 \text{ kJ/mol}$$

This ΔH value tells us that when 1 mole of methane is burned in oxygen, 728 kJ of energy are released. This value is called the **enthalpy change of combustion** of methane.

❓ Worked example

Look at the reaction pathway diagrams below.

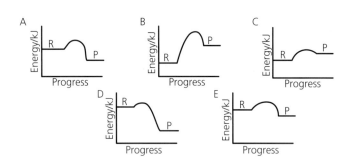

Which of the diagram(s) shows:

a an exothermic reaction?

The exothermic reactions will show the products at a lower energy than the reactants, showing that energy has been given out when the reaction occurs. So the exothermic reactions are A, D and E.

b an endothermic reaction?

The endothermic reactions will have the products at a higher energy than the reactants, showing that energy is being taken in, or added, when the reaction occurs. So the endothermic reactions are B and C.

c the most endothermic reaction?

The two endothermic reactions were reactions B and C. The most endothermic reaction will have the biggest gap between the energy of the reactants and products. In this example that is reaction B.

d the reaction with the highest activation energy, E_a?

To find which reaction has the highest activation energy draw a line across from the reactant line and then up to the top of the curve. The reaction which has the highest activation energy is the one with the longest vertical line to the top of the curve. In this example that will be reaction B.

? Worked example

Look at the two reactions below, which have been given with their ΔH values.

$$C + H_2O \rightarrow H_2 + CO;\ \Delta H = +131 \text{ kJ/mol}$$
$$C + O_2 \rightarrow CO_2;\ \Delta H = -393 \text{ kJ/mol}$$

For each of these reactions, draw and label a reaction pathway diagram and label on each diagram the reactants, the products, the enthalpy change of reaction, ΔH, and the activation energy, E_a.

The first reaction is an endothermic reaction, so we need to draw a reaction pathway diagram which has the products at a higher energy than the reactants, showing that energy is being gained (or added) when the reaction occurs.

You should be able to label the ΔH value and place the symbols for the reactants and products on their lines. No information was given about the activation energy in this question so the shape and height of the curve is not specific.

The second reaction is an exothermic reaction, so we need to draw a reaction pathway diagram which has the products at a lower energy than the reactants, showing energy is being given out (lost) when the reaction occurs.

You should be able to label the ΔH value and place the symbols for the reactants and products on their lines. No information was given about the activation energy in this question so the shape and height of the curve is not specific.

? Worked example

Using the bond energies in Table 6.1, calculate the enthalpy of reaction, ΔH, when the hydrocarbon ethane, C_2H_6, is completely combusted.

$$\text{H}-\overset{\overset{\text{H}}{|}}{\underset{\underset{\text{H}}{|}}{\text{C}}}-\overset{\overset{\text{H}}{|}}{\underset{\underset{\text{H}}{|}}{\text{C}}}-\text{H} + 3\tfrac{1}{2}\text{O}=\text{O} \rightarrow 2\text{O}=\text{C}=\text{O} + 3\text{H}-\overset{\text{O}}{}-\text{H}$$

The first step is to work out the total amount of energy needed to break the bonds in the reactants. The reactants are 1 mole of ethane and 3½ moles of oxygen gas.

Bond breaking:

Breaking 6 C–H bonds requires 6 × 435 = 2610 kJ

Breaking one C–C bond requires 347 kJ

Breaking 3½ moles of O=O bonds requires 3½ × 497 = 1739.5 kJ

Total = 4696.5 kJ

The second step is to work out the total amount of energy which is released when the bonds in the products are formed. The products are 2 moles of carbon dioxide and 3 moles of water.

Bond making:

Making 4 C=O bonds releases 4 × 803 = 3212 kJ

Making 6 H–O bonds releases 6 × 464 = 2784 kJ

Total = 5996 kJ

enthalpy of reaction = energy required to break the bonds – energy given off when bonds are made

= 4696.5 – 5996

= –1299.5 kJ

Going further

Enthalpy change of neutralisation (molar heat of neutralisation)

This is the enthalpy change that takes place when 1 mole of hydrogen ions ($H^+(aq)$) is neutralised.

$$H^+(aq) + OH^-(aq) \rightarrow H_2O(l); \Delta H = -57 \text{ kJ/mol}$$

This process occurs in the titration of an alkali by an acid to produce a neutral solution (Chapter 8, p. 128).

Practical skills

Experiment to find the enthalpy change of reaction

For safe experiments/demonstrations which are related to this chapter, please refer to the *Cambridge IGCSE Chemistry Practical Skills Workbook*, which is also part of this series.

Safety

- Eye protection must be worn.
- Take care with copper(II) sulfate solution: $\leq 1.0 \text{ mol/dm}^3$ is an irritant and corrosive. Disposable gloves should be worn when measuring out.

A student uses a simple experiment to find the energy given off when a chemical reaction occurs (the enthalpy change of reaction, ΔH) in the reaction between zinc metal and copper(II) sulfate. When zinc reacts with copper(II) sulfate, the following reaction occurs:

zinc + copper(II) sulfate → zinc sulfate + copper

$$Zn(s) + CuSO_4(aq) \rightarrow ZnSO_4(aq) + Cu(s)$$

They place an expanded polystyrene cup into a beaker and pour into the expanded polystyrene cup 25 cm³ of 1 mol/dm³ copper sulfate solution and records the temperature of the copper(II) sulfate solution using a thermometer. They then add 5 g of zinc powder, which is in excess. The student stirred the mixture carefully with the thermometer and recorded the highest temperature reached.

Find the change in temperature and then use the following relationship to find the thermal energy transferred during the reaction:

thermal energy transferred during the reaction
= mass of × 4.2 J/g/°C × change in
  solution/g   temperature/°C

1 Why is an expanded polystyrene cup used for the reaction to occur in?
2 What is the purpose of placing the expanded polystyrene cup into the beaker?
3 Why is the mixture stirred after the zinc is added?
4 Use these questions to identify improvements to this experiment.
   a Think of possible sources of errors in this experiment.
   b Give some improvements which will lessen these errors.

Sample data:

Initial temperature of the copper(II) sulfate solution = 19.5 °C

Maximum temperature reached = 26.9 °C

5 Use the relationship above to determine the thermal energy transferred in joules. (Assume that the density of the copper(II) sulfate solution is 1 g/cm³.)
6 Calculate the number of moles of copper(II) sulfate used in the reaction.

You now know the amount of energy given off when the number of moles you have calculated in question **6** have reacted.

7 Calculate the energy that would have been given off if 1 mole of copper(II) sulfate had been used.
8 Convert the energy given off into kJ and write down the value of the enthalpy change of reaction, ΔH, you have calculated.

Test yourself

11 Use the bond energy data given in Table 6.1 to answer the following questions.
   a Calculate the enthalpy change per mole for the combustion of ethanol, a fuel added to petrol in some countries.
   b Draw a reaction pathway diagram to represent this combustion process.
   c How does this compare with the enthalpy of combustion of heptane (C_7H_{16}), a major component of petrol, of −4853 kJ/mol?
   d How much energy is released per gram of ethanol and heptane burned?
12 How much energy is released if:
   a 0.5 mole of methane is burned?
   b 5 moles of methane are burned?
   c 4 g of methane are burned?
      (A_r: C = 12; H = 1)
13 How much energy is released if:
   a 2 moles of hydrogen ions are neutralised?
   b 0.25 moles of hydrogen ions are neutralised?
   c 1 mole of sulfuric acid is completely neutralised?

Revision checklist

After studying Chapter 6 you should be able to:
✓ Describe the fractionation of petroleum to produce different fractions.
✓ Describe the changing properties of the fractions as they are produced from the fractionating column.
✓ Name uses for the fractions produced.
✓ Explain what exothermic and endothermic reactions are.

___
✓ Explain why energy is given out or taken in during reactions in terms of the bond breaking and bond making which occurs when the reaction happens.
___

✓ Interpret and draw reaction pathway diagrams.

___
✓ Define activation energy.
✓ Calculate the enthalpy change of reaction by using bond energy data.
___

✓ Identify coal, natural gas and petroleum as fossil fuels, and that natural gas contains mainly methane.
✓ State that hydrocarbons are compounds that contain only the elements hydrogen and carbon.
✓ State that petroleum is a mixture of different hydrocarbon molecules.

Exam-style questions

1 a State which of the following processes is endothermic and which is exothermic.
   i   the breaking of a chemical bond [1]
   ii the forming of a chemical bond [1]
   b The table below shows the bond energy data for a series of covalent bonds.

Bond	Bond energy/kJ/ mol
C—H	435
O=O	497
C=O	803
H—O	464
C—C	347
C—O	358

   i   Use the information given in the table to calculate the overall enthalpy change for the combustion of ethanol producing carbon dioxide and water. [3]
   ii Is the process in b i endothermic or exothermic? [1]

2 Petroleum is a mixture of hydrocarbons. The refining of petroleum produces fractions which are more useful to us than petroleum itself. Each fraction is composed of hydrocarbons which have boiling points within a specific range of temperature. The separation is carried out in a fractionating column, as shown below.

   a Which separation technique is used to separate the fractions? [1]

   b Name each of the fractions A to H and give a use for each. [8]
   c Why do the fractions come from the fractionating column in this order? [3]
   d What is the connection between your answer to c and the size of the molecules in each fraction? [1]
   e Which of the fractions will be the most flammable? [1]

3 'Propagas' is used in some central heating systems where natural gas is not available. It burns according to the following equation:
$$C_3H_8(g) + 5O_2(g) \rightarrow 3CO_2(g) + 4H_2O(l);$$
$$\Delta H = -2220 \text{ kJ/mol}$$

   a What are the chemical names for propagas and natural gas?
   b Would you expect the thermal energy released per mole of propagas burned to be greater than that for natural gas? Explain your answer.
   c What is propagas obtained from?
   d Calculate:
      i   the mass of propagas required to produce 5550 kJ of energy
      ii the thermal energy produced by burning 0.5 mole of propagas
      iii the thermal energy produced by burning 11 g of propagas
      iv the thermal energy produced by burning 2000 dm³ of propagas
          (A_r: H = 1; C = 12; O = 16. 1 mole of any gas occupies 24 dm³ at room temperature and pressure.)

4 This question is about endothermic and exothermic reactions.
   a Explain the meaning of the terms endothermic and exothermic. [2]

   b i Draw a reaction pathway diagram for the reaction: [4]
$$NaOH(aq) + HCl(aq) \rightarrow NaCl(aq) + H_2O(l);$$
$$\Delta H = -57 \text{ kJ/mol}$$

      ii Is this reaction endothermic or exothermic? [1]
      iii Calculate the enthalpy change associated with this reaction if 2 moles of sodium hydroxide were neutralised by excess hydrochloric acid. [1]

c i Draw a reaction pathway diagram for the reaction: [4]

$$2H_2O(l) \rightarrow 2H_2(g) + O_2(g); \Delta H = +575 \text{ kJ/mol}$$

ii Is this reaction endothermic or exothermic? [1]

iii Calculate the enthalpy change for this reaction if only 9 g of water was converted into hydrogen and oxygen. [2]

5 The following results were obtained from an experiment carried out to measure the enthalpy change of combustion of ethanol. The experiment involved heating a known volume of water with the flame from an ethanol burner.

Thermometer

Metal calorimeter

Clamp

Water

Heat

Draught shield

The burner was weighed initially and after the desired temperature rise had been obtained.

Volume of water in glass beaker = 200 cm³

Mass of ethanol burner at start = 85.3 g

Mass of ethanol burner at end = 84.8 g

Temperature rise of water = 12 °C

(Density of water = 1 g/cm³)

Thermal energy transferred to water = mass of water/g × 4.2 J/g/°C × temperature rise/°C

a Calculate the mass of ethanol burned. [1]

b Calculate the amount of energy produced, in joules, in this experiment by the ethanol burning. [2]

c Convert your answer to **b** into kilojoules. [1]

d Calculate the amount of energy produced by 1 g of ethanol burning. [1]

e What is the molar mass of ethanol (C₂H₅OH)? [1]
(A_r: H = 1; C = 12; O = 16)

f How much energy would be produced if 1 mole of ethanol had been burned? (This is the enthalpy change of combustion of ethanol.) [1]

g Compare your value with the actual value of 1371 kJ/mol and suggest two reasons for the difference in values. [3]

h Write a balanced chemical equation to represent the combustion of ethanol. [1]

6 The following results were obtained from a neutralisation reaction between 1 mol/dm³ hydrochloric acid and 1 mol/dm³ sodium hydroxide. This experiment was carried out to measure the enthalpy change of neutralisation of hydrochloric acid. The temperature rise which occurred during the reaction was recorded.

Volume of sodium hydroxide used = 50 cm³

Volume of acid used = 50 cm³

Temperature rise = 5°C

(Density of water = 1 g/cm³)

Thermal energy transferred to water = mass of water/g × 4.2 J/g/°C × temperature rise/°C

a Write a balanced chemical equation for the reaction. [1]

b What mass of solution was warmed during the reaction? [1]

c How much thermal energy was produced during the reaction? [1]

d How many moles of hydrochloric acid were involved in the reaction? [2]

e How much thermal energy would be produced if 1 mole of hydrochloric acid had reacted? (This is the enthalpy change of neutralisation of hydrochloric acid.) [2]

f The enthalpy change of neutralisation of hydrochloric acid is −57 kJ/mol. Suggest two reasons why there is a difference between this and your calculated value. [2]

7 Chemical reactions

FOCUS POINTS
- ★ How does the rate of a reaction vary?
- ★ Why is the rate of a chemical reaction important?
- ★ What is a reversible reaction?

In this chapter, we will look at how we are able to alter the rate at which chemical reactions occur. Industry needs to produce chemical products in the shortest possible time, so they are made efficiently and increase profits. We know that all chemical reactions go faster when heated to a higher temperature, for example, an egg cooks faster at higher temperatures, but why does this happen?

Industry also wants to make more product from a reaction. In any chemical reaction, reactants will make products eventually, so how can there be more product? We will introduce the fact that not all reactions simply produce products from reactants: some reactions are reversible and, once the product has been formed, it may breakdown to produce reactants again.

By the end of this chapter, you should be able to fully explain how the rates of chemical reactions can be altered and how reversible reactions can be manipulated to produce products efficiently.

7.1 Reactions

A physical change does not result in the formation of a new substance. It simply alters the state of the substance involved. For example, when ice melts to form water, it is a physical change because the water has not changed: only its state has changed from a solid to a liquid. This is also the case if we boil water to produce steam.

A chemical change, however, results in the formation of something new. An everyday example of this is frying an egg. The chemicals in the fried egg have undergone chemical reactions forming new substances, and the fried egg cannot be changed back into the egg you started off with. This is the main difference between a physical and chemical change.

In a physical change, it is possible to reverse the change, and the change can be temporary. For example, the water from melted ice can be frozen again to form ice. In a chemical change, this cannot happen. For all chemical reactions, a chemical change occurs and this is a permanent change.

Figure 7.1 shows some slow and fast reactions. The two photographs at the top show examples of slow reactions. The ripening of apples takes place over a number of weeks, and the making and maturing of cheese may take months. The burning of solid fuels, such as coal, can be said to involve chemical reactions taking place at a medium speed or rate. A fast reaction is an explosion (bottom right), where the chemicals inside explosives, such as TNT, react very rapidly in reactions which are over in seconds or fractions of seconds.

As new techniques have been developed, the processes used within the chemical industry have become more complex. Therefore, chemists and chemical engineers have increasingly looked for ways to control the rates at which chemical reactions take place. In doing so, they have discovered that there are five main ways in which you can alter the rate of a chemical reaction. These ideas are not only incredibly useful to industry but can also be applied to reactions in the school laboratory.

▲ **Figure 7.1** Some slow (ripening fruit and cheese making), medium (coal fire) and fast (explosion) reactions

7.2 Factors that affect the rate of a reaction

The five main ways to alter the rate of a chemical reaction are:

» changing the concentration of solutions
» changing the pressure of gases
» changing the surface area of solids
» changing the temperature
» adding or removing a catalyst including enzymes.

Collision theory

For a chemical reaction to occur, reactant particles need to collide with one another. For products to be formed, the collision has to have a certain minimum amount of energy associated with it,

and not every collision results in the formation of products. This minimum amount of energy is known as the **activation energy**, E_a (Figure 7.2). Collisions which result in the formation of products are known as successful collisions.

▲ **Figure 7.2** Energy level diagram showing activation energy

Surface area

Limestone (calcium carbonate, $CaCO_3$) is a substance which can be used to neutralise soil acidity. Powdered limestone is used as it neutralises the acidity faster than if lumps of limestone are used. Why do you think this is the case?

In the laboratory, the reaction between acid and limestone in the form of lumps or powder can be observed in a simple test-tube experiment. Figure 7.3 shows the reaction between dilute hydrochloric acid and limestone in lump and powdered form.

$$\text{hydrochloric} + \text{calcium} \rightarrow \text{calcium} + \text{carbon} + \text{water}$$
$$\text{acid} \qquad \text{carbonate} \qquad \text{chloride} \quad \text{dioxide}$$

$$2HCl(aq) + CaCO_3(s) \rightarrow CaCl_2(aq) + CO_2(g) + H_2O(l)$$

▲ **Figure 7.3** The powdered limestone (left) reacts faster with the acid than the limestone in the form of lumps

The rates at which the two reactions occur can be found by measuring either:

» the volume of the carbon dioxide gas which is produced
» the loss in mass of the reaction mixture with time.

These two methods are generally used for measuring the rate of reaction for processes involving the formation of a gas as one of the products.

The apparatus shown in Figure 7.4 is used to measure the loss in mass of the reaction mixture. The mass of the conical flask plus the reaction

mixture is measured at regular intervals. The total loss in mass is calculated for each reading of the balance, and this is plotted against time. Some sample results from experiments of this kind have been plotted in Figure 7.5.

The reaction between hydrochloric acid and limestone is generally at its fastest in the first minute. This is indicated by the slopes of the curves during this time. The steeper the slope, the faster the rate of reaction. You can see from the two curves in Figure 7.5 that the rate of reaction is greater with the powdered limestone than the lump form.

▲ **Figure 7.4** After 60 seconds the mass has fallen by 1.24 g

If the surface area of a reactant is increased, more particles are exposed to the other reactant, so the rate of a chemical reaction can be raised by increasing the surface area of a solid reactant.

The surface area has been increased by powdering the limestone (Figure 7.6). The acid particles now have an increased amount of surface of limestone with which to collide. The products of a reaction are formed when collisions occur between reactant particles.

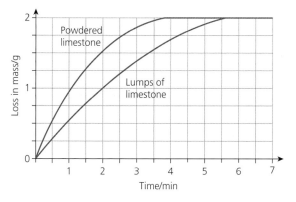

▲ **Figure 7.5** Sample results for the limestone/acid experiment

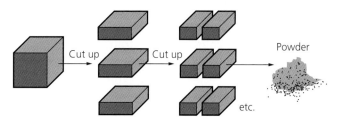

▲ **Figure 7.6** A powder has a larger surface area

Practical skills

Experiment to show the effect of surface area on reaction rate

For safe experiments/demonstrations which are related to this chapter, please refer to the *Cambridge IGCSE Chemistry Practical Skills Workbook*, which is also part of this series.

Safety

- Eye protection must be worn.

A student took a lump of limestone (calcium carbonate) and reacted it with an excess of dilute hydrochloric acid in a conical flask which had a delivery tube attached. The gas produced was collected in an inverted burette filled with water. The experiment was repeated using smaller fragments of limestone of the same mass. The table on the right gives the results of the experiments.

1 Draw the apparatus which could be used in the experiment above.
2 Plot a graph of volume of carbon dioxide collected (*y*-axis) against time (*x*-axis) for both experiments. Use the same axes.
3 a Which of the experiments is the fastest?
  b How can you tell this from the graph?
  c Use ideas about particles to explain your answer.

4 a After how many seconds was the experiment using the single lump of limestone complete?
  b How can you tell this from the graph?
5 What other factors must be kept constant during this experiment to show the effect of surface area on the rate of reaction?
6 Can you think of any sources of errors with the method used?

Time/secs	A lump of limestone Volume of CO_2/cm³	Smaller fragments of limestone Volume of CO_2/cm³
0	0	0
10	12	25
20	25	42
30	36	53
40	46	62
50	54	66
60	61	68
70	65	68
80	68	68
90	68	68
100	68	68

An increase in the surface area of a solid reactant results in an increase in the number of collisions, and this results in an increase in the number of successful collisions. Therefore, the increase in surface area of the limestone increases the rate of reaction.

In certain industries the large surface area of fine powders and dusts can be a problem. For example, there is a risk of explosion in flourmills and mines, where the large surface area of the flour or coal dust can – and has – resulted in explosions through a reaction with oxygen gas in the air when a spark has been created by machinery or the workforce (Figure 7.7). On 26 September 1988, two silos containing wheat exploded at the Jamaica Flour Mills Plant in Kingston, Jamaica, killing three workers, as a result of fine dust exploding.

▲ **Figure 7.7** The dust created by this cement plant is a potential hazard

Test yourself

1 What apparatus would you use to measure the rate of reaction of limestone with dilute hydrochloric acid by measuring the volume of carbon dioxide produced?

2 The following results were obtained from an experiment of the type you were asked to design in question **1**.

Time/ min	0	0.5	1.0	1.5	2.0	2.5	3.0	3.5	4.0	4.5	5.0
Total volume of CO_2 gas/cm^3	0	15	24	28	31	33	35	35	35	35	35

  a Plot a graph of the total volume of CO_2 against time.
  b At which point is the rate of reaction fastest?
  c What volume of CO_2 was produced after 1 minute 15 seconds?
  d How long did it take to produce $30\,cm^3$ of CO_2?

Concentration

A yellow precipitate is produced in the reaction between sodium thiosulfate and hydrochloric acid.

sodium + hydrochloric → sodium + sulfur + sulfur + water
thiosulfate   acid   chloride   dioxide

$$Na_2S_2O_3(aq) + 2HCl(aq) \rightarrow 2NaCl(aq) + S(s) + SO_2(g) + H_2O(l)$$

The rate of this reaction can be followed by recording the time taken for a given amount of the yellow sulfur to be precipitated. This can be done by placing a conical flask containing the reaction mixture on to a cross on a piece of paper (Figure 7.8). As the precipitate of sulfur forms, the cross is obscured and finally disappears from view. The time taken for this to occur is a measure of the rate of this reaction. To obtain sufficient information about the effect of changing the concentration of the reactants, several experiments of this type must be carried out, using different concentrations of sodium thiosulfate or hydrochloric acid. If you are

▲ **Figure 7.8** The precipitate of sulfur obscures the cross

considering carrying out this practical, take care as SO_2 is released, which can cause breathing issues, especially for asthmatics.

Some sample results of experiments of this kind have been plotted in Figure 7.9. You can see from the graph that when the most concentrated sodium thiosulfate solution was used, the reaction was at its fastest. This is shown by the shortest time taken for the cross to be obscured.

▲ **Figure 7.9** Sample data for the sodium thiosulfate/ acid experiment at different concentrations of sodium thiosulfate

From the data shown in Figure 7.9 it is possible to produce a different graph which directly shows the rate of the reaction against concentration rather than the time taken for the reaction to occur against concentration. To do this, the times can be converted to a rate using:

$$rate = \frac{1}{reaction\ time\ (s)}$$

This would give the graph shown in Figure 7.10.

▲ **Figure 7.10** Graph to show the rate of reaction against concentration

The products of a reaction are formed as a result of the collisions between reactant particles. In a more concentrated solution there are more particles, which leads to more frequent collisions between reactant particles. The more often the particles collide, the greater the chance they have of having sufficient energy to overcome the activation energy of the reaction, and so it is more likely that a successful collision occurs. This means that the rate of a chemical reaction will increase if the concentration of reactants is increased, because there are more particles per unit volume.

Pressure of gases

In reactions involving only gases, an increase in the overall pressure at which the reaction is carried out increases the rate of the reaction. The increase in pressure results in the gas particles being pushed closer together, increasing the number of particles per unit volume. This means that they collide more frequently and so the rate of reaction increases.

Temperature

Why do you think food is stored in a refrigerator? The reason is that the rate of decay of food is slower at lower temperatures. This is a general feature of the majority of chemical processes.

The reaction between sodium thiosulfate and hydrochloric acid can be used to study the effect of temperature on the rate of a reaction. Figure 7.11 shows some sample results of experiments with sodium thiosulfate and hydrochloric acid (at fixed concentrations) carried out at different temperatures. You can see from the graph that the rate of the reaction is fastest at high temperatures.

▲ **Figure 7.11** Sample data for the sodium thiosulfate/acid experiment at different temperatures

As the temperature increases, the reactant particles increase their kinetic energy. Some of the particles will have an energy greater than the activation energy, E_a, and they move faster. The faster movement results in more frequent collisions between the particles. Some of the extra collisions that result from the temperature increase will be successful collisions. This causes the **reaction rate** to increase.

▲ **Figure 7.12** Sample data for differing amounts of MnO_2 catalyst

Test yourself

3 Explain why potatoes cooked in oil cook faster than those cooked in water.
4 Devise an experiment to study the effect of temperature on the reaction between magnesium and hydrochloric acid.
5 Explain why food cooks faster in a pressure cooker.

Catalysts

Over 90% of industrial processes use **catalysts**. A catalyst is a substance which can alter the rate of a reaction without being chemically changed itself. In the laboratory, the effect of a catalyst can be observed using the decomposition of hydrogen peroxide as an example.

hydrogen peroxide $\rightarrow$ water $\;$ + oxygen

$$2H_2O_2(aq) \quad \rightarrow 2H_2O(l) + \;\; O_2(g)$$

The rate of decomposition at room temperature is very slow. There are substances that can speed up this reaction, one being manganese(IV) oxide. When black manganese(IV) oxide powder is added to hydrogen peroxide solution, oxygen is produced rapidly. The rate at which this occurs can be seen by measuring the volume of oxygen gas produced with time.

Some results from experiments of this type have been plotted in Figure 7.12. At the end of the reaction, the manganese(IV) oxide can be filtered off and used again. The reaction can proceed even faster by increasing the amount and surface area of the catalyst. This is because the activity of a catalyst involves its surface. Note that, in gaseous reactions, if dirt or impurities are present on the surface of the catalyst, it will not act as efficiently;

it is said to have been 'poisoned'. Therefore, the gaseous reactants must be pure.

Chemists have found that:

» a small amount of catalyst will produce a large amount of chemical change
» catalysts remain unchanged chemically after a reaction has taken place, but they can change physically. For example, a finer manganese(IV) oxide powder is left behind after the decomposition of hydrogen peroxide
» catalysts are specific to a particular chemical reaction.

Some examples of chemical processes and the catalysts used are shown in Table 7.1.

▼ **Table 7.1** Examples of catalysts

Process	Catalyst
Haber process – for the manufacture of ammonia	Iron
Contact process – for the manufacture of sulfuric acid	Vanadium(V) oxide
Oxidation of ammonia to give nitric acid	Platinum
Hydrogenation of unsaturated oils to form fats in the manufacture of margarines	Nickel

Key definition

A **catalyst** is a substance which alters the rate of a chemical reaction and is unchanged at the end of the reaction.

A catalyst increases the rate of a chemical reaction by providing an alternative reaction path which has a lower activation energy, E_a. A catalyst does not increase the number of collisions between the reactant particles but it causes more of the collisions to become successful collisions, so increasing the rate of the reaction.

If the activation energy is lowered by using a catalyst then, on collision, more particles will go on to produce products at a given temperature (Figure 7.13).

▲ **Figure 7.13** A reaction pathway diagram showing activation energy, with and without a catalyst

▲ **Figure 7.14** The enzyme molecules (red, pink, green and blue) lock on exactly to a particular reactant molecule (yellow). Once the enzyme is locked on, the reactant molecule breaks up into pieces

For example, hydrogen peroxide is produced within our bodies. However, it is extremely damaging and must be decomposed very rapidly. Catalase is the enzyme which converts hydrogen peroxide into harmless water and oxygen within our livers:

$$\text{hydrogen peroxide} \xrightarrow{\text{catalase}} \text{water} + \text{oxygen}$$
$$2\text{H}_2\text{O}_2(\text{aq}) \xrightarrow{\text{catalase}} 2\text{H}_2\text{O}(\text{l}) + \text{O}_2(\text{g})$$

Although many chemical catalysts can work under various conditions of temperature and pressure, as well as alkalinity or acidity, biological catalysts operate only under very particular conditions. For example, they operate over a very narrow temperature range and, if the temperature becomes too high, they become inoperative. At temperatures above about 45°C, they denature. This means that the specific shape of the active site of the enzyme molecule changes due to the breaking of bonds. This means that the reactant molecules are no longer able to fit into the active site.

A huge multimillion-pound industry has grown up around the use of enzymes to produce new materials. Biological washing powders (Figure 7.15) contain enzymes to break down stains such as sweat, blood and egg, and they do this at the relatively low temperature of 40°C. This reduces energy costs, because the washing water does not need to be heated as much.

> **Test yourself**
>
> **6** Using a catalysed reaction of your choice, devise an experiment to follow the progress of the reaction and determine how effective the catalyst is.

7.3 Enzymes

Enzymes are protein molecules produced in living cells. They are catalysts which speed up hundreds of different chemical reactions taking place inside living cells. These biological catalysts are very specific in that each chemical reaction has a different enzyme catalyst. There are literally hundreds of different kinds of enzyme. Enzymes all have an active site. The active site is a particular shape and locks into a corresponding shape in a reactant molecule. When this has happened, the enzyme can work to break up the reactant (Figure 7.14).

▲ **Figure 7.15** These biological washing powders contain enzymes

There have been problems associated with the early biological washing powders, and many manufacturers still place warnings on their packets, indicating that the powder contains enzymes which may cause skin rashes. Some customers suffered from skin rashes because they were allergic to the enzymes. This problem has been overcome to a certain extent by advising that extra rinsing is required.

➡ Going further

Other industrial processes also make use of enzymes.

- In the manufacture of baby foods, enzymes called proteases are used to 'pre-digest' the protein part of the baby food.
- In the production of yoghurt, enzymes are added to partially digested milk protein, allowing it to be more easily absorbed by the body.
- In cheese making, enzymes are added to coagulate milk and to separate it into solids (curds) and liquid (whey).

In industry, enzymes are used to bring about reactions at normal temperatures and pressures that would otherwise require expensive conditions and equipment. Successful processes using enzymes need to ensure that:

- the enzyme is able to function for long periods of time by optimising the environment
- the enzyme is kept in place by trapping it on the surface of an inert solid (some processes immobilise the enzymes when the process is complete)
- continuous processes occur rather than batch processes.

▶ Test yourself

7 When using biological washing powders what factors have to be taken into consideration?
8 Enzymes in yeast are used in the fermentation of glucose. Why, when the temperature is raised to 45°C, is very little ethanol actually produced compared with the amount formed at room temperature?

Rate of reaction is very important in the chemical industry. There are, however, many other factors which the industry needs to take into account as they manufacture chemicals such as safety and cost. In the next section we will look at the industrial manufacture of a very important chemical called ammonia.

7.4 Reversible reactions and equilibrium

Most chemical reactions simply change reactants into products. For example, carbon reacts with oxygen to produce carbon dioxide, but the carbon dioxide produced does not easily change back to give carbon and oxygen again. There are, however, some chemical reactions which are reversible. These are reactions in which reactants form products which can then react further to produce more reactants, which reverse the process and produce the original reactants. The resulting mixture may contain reactants and products, and their proportions can be changed by altering the reaction conditions.

Some of these reactions produce important industrial chemicals, such as ammonia and sulfuric acid. We will look at the industrial production of ammonia later in this chapter but first let us look at some simpler examples of reversible reactions.

Copper(II) sulfate is available in two different forms: **hydrated** copper(II) sulfate, which is a blue solid, and **anhydrous** copper(II) sulfate which is a white solid. If hydrated copper(II) sulfate is heated, water is removed and the colour changes from blue to white, as anhydrous copper(II) sulfate is formed. If water is then added to the white solid, it changes back to a blue colour as the hydrated copper(II) sulfate is reformed (Figure 7.16).

▲ **Figure 7.16** Anhydrous copper(II) sulfate changes from white to blue when water is added to it

This is an example of a reversible reaction:

hydrated copper(II) $\rightleftharpoons$ anhydrous copper(II) + water
    sulfate   sulfate

  blue solid $\rightleftharpoons$ white solid

We can show this reaction as being **reversible** by using the $\rightleftharpoons$ symbol.

> **Key definition**
> Some chemical reactions are **reversible** as shown by the $\rightleftharpoons$ symbol.

Another example is cobalt(II) chloride, which is a blue solid in its anhydrous form and pink in its hydrated form. Adding water to anhydrous cobalt(II) chloride changes the colour from blue to a pink solid as it produces hydrated cobalt(II) chloride. If hydrated cobalt(II) chloride is heated the colour changes from pink back to blue.

  anhydrous cobalt(II) + water $\rightleftharpoons$ hydrated cobalt(II)
    chloride   chloride

    blue solid   pink solid

You will learn more about copper(II) sulfate in Chapter 8 (p. 123). Both examples shown above can be used as a chemical test for the presence of water.

In summary, heating a hydrated compound will remove the water and form the anhydrous compound. Adding some water to an anhydrous compound will produce the hydrated compound.

7.5 Ammonia – an important nitrogen-containing chemical

Nitrogen from the air is used to manufacture ammonia, a very important bulk chemical. A bulk chemical is one that is used in large quantities, across a range of uses. The major process used for making ammonia is the **Haber process**. This process was developed by the German scientist Fritz Haber in 1913. He was awarded a Nobel Prize in 1918 for his work. The process involves reacting nitrogen and hydrogen. Ammonia has many uses including the manufacture of explosives, nitric acid and fertilisers such as ammonium nitrate.

Obtaining nitrogen

The nitrogen needed in the Haber process is obtained from the **atmosphere** by fractional distillation of liquid air (Chapter 12, p. 229).

Obtaining hydrogen

The hydrogen needed in the Haber process is obtained from the reaction between methane and steam.

  methane + steam $\rightleftharpoons$ hydrogen + carbon monoxide

  $CH_4(g)$ + $H_2O(g) \rightleftharpoons 3H_2(g)$ + $CO(g)$

This process is known as steam re-forming and is reversible.

In any reversible reaction, the reaction shown going from left to right in the equation is known as the forward reaction. The reaction shown going from right to left in the equation is known as the reverse reaction. Special conditions are employed to ensure that the reaction proceeds as the forward reaction (to the right in the equation), producing hydrogen and carbon monoxide. The process is carried out at a temperature of 750°C, at a pressure of 3000 kPa (30 atmospheres) with a catalyst of nickel. These conditions enable the maximum amount of hydrogen to be produced at an economic cost.

The carbon monoxide produced is then allowed to reduce some of the unreacted steam to produce more hydrogen gas.

carbon monoxide + steam $\rightleftharpoons$ hydrogen + carbon dioxide

$$CO(g) + H_2O(g) \rightleftharpoons H_2(g) + CO_2(g)$$

Other ways of producing hydrogen for the Haber process do exist, usually using hydrocarbons, but steam reforming is the most commonly used method.

> **Key definition**
>
> The source of hydrogen used in the Haber process is methane, and nitrogen gas is obtained from the air.

Making ammonia

In the Haber process, nitrogen and hydrogen in the correct proportions (1:3) are pressurised to approximately 20 000 kPa (200 atmospheres) and passed over a catalyst of freshly produced, finely divided iron at a temperature of 450°C. The reaction in the Haber process is:

nitrogen + hydrogen $\rightleftharpoons$ ammonia

$$N_2(g) + 3H_2(g) \rightleftharpoons 2NH_3(g) \quad \Delta H = -92\,kJ/mol$$

The reaction is exothermic.

> **Key definition**
>
> The equation for the production of ammonia in the Haber process is $N_2 + 3H_2 \rightleftharpoons 2NH_3$.
>
> The typical conditions used in the Haber process are 450°C, 20 000 kPa and an iron catalyst.

The industrial conditions employed ensure that sufficient ammonia is produced at a fast enough rate. Under these conditions, the gas mixture leaving the reaction vessel contains about 15% ammonia, which is removed by cooling and condensing it as a liquid. The unreacted nitrogen and hydrogen are re-circulated into the reaction vessel to react together once more to produce further quantities of ammonia.

Equilibrium

The 15% of ammonia produced does not seem a great deal. The reason for this is the reversible nature of the reaction. Once the ammonia is made from nitrogen and hydrogen, it decomposes to produce nitrogen and hydrogen. There comes a point when the rate at which the nitrogen and hydrogen react to produce ammonia is equal to the rate at which the

ammonia decomposes. The concentrations of nitrogen, hydrogen and ammonia in the reaction vessel do not change, provided that the reaction vessel is a closed system: one in which none of the gases can enter or leave. This situation is called a chemical equilibrium. Because the processes continue to happen, the equilibrium is said to be dynamic. A **reversible reaction**, in a closed system, is at equilibrium when the rate of the forward and reverse reactions are equal and the concentrations of the reactants and products remain constant. The conditions used ensure that the ammonia is made economically. Figure 7.17 shows how the percentage of ammonia produced varies with the use of different temperatures and pressures.

> **Key definitions**
>
> A **reversible reaction** in a closed system is at equilibrium when the rate of the forward reaction is equal to the rate of the reverse reaction, and the concentrations of reactants and products are no longer changing.

▲ **Figure 7.17** Yields from the Haber process

You will notice from Figure 7.17 that the higher the pressure and the lower the temperature used, the more ammonia is produced. In fact, the position of equilibrium is affected by each of the following:

» changing temperature
» changing pressure
» changing concentration
» using a catalyst.

Relationships such as these were initially observed by Henri Le Chatelier, a French scientist, in 1888 (Figure 7.18). He noticed that if the pressure was increased in reactions involving gases, the reaction which produced the fewest molecules of gas was favoured. If you look at the reaction for the Haber process, you will see that, going from left to right, the number of molecules of gas goes from four to two. This is why the Haber process is carried out at high pressures. We say that the position of equilibrium has been moved to the right by using the higher pressure. Le Chatelier also noticed that reactions which were exothermic produced more products if the temperature was low. Using a lower temperature favours the exothermic process. In the Haber process, this would move the position of equilibrium to the right. Indeed, if the Haber process is carried out at room temperature, you get a higher percentage of ammonia. However, in practice the rate of the reaction is lowered too much and the ammonia is not produced quickly enough for the process to be economical. An **optimum temperature** is used to produce enough ammonia at an acceptable rate.

A catalyst is used in almost all industrial processes to produce the product more quickly, and in the Haber process an iron catalyst is used. The catalyst increases the rate of both the forward and reverse reactions to the same extent, so the position is not affected: its use simply means the product is produced more quickly.

The concentration of nitrogen is often increased as this will also favour the forward reaction, increasing the yield of ammonia.

It should be noted, however, that the high pressure used is very expensive and is a major safety concern so alternative, less expensive routes involving biotechnology are currently being sought. Worldwide, in excess of 175 million tonnes of ammonia are produced by the Haber process each year.

> ### ▶ Test yourself
>
> 9 What problems do the builders of a chemical plant designed to produce ammonia have to consider when they start to build such a plant?
> 10 What problems are associated with building a plant which uses such high pressures as those required in the Haber process?

7.6 Industrial manufacture of sulfuric acid – the Contact process

Sulfuric acid is probably the most important industrial chemical, and the quantity of it produced by a country has been linked with the economic stability of the country. Many million tonnes of sulfuric acid are produced worldwide each year. It is used mainly as the raw material for the production of substances such as detergents, paints, fertilisers and organic compounds including plastics.

The process by which sulfuric acid is produced is known as the **Contact process**.

The process has the following stages.

Sulfur dioxide is first produced by burning sulfur with an excess of oxygen in the air.

sulfur + oxygen → sulfur dioxide

$$S(s) + O_2(g) \rightarrow SO_2(g)$$

▲ **Figure 7.18** Henri Le Chatelier (1850–1936)

The sulfur dioxide can also be obtained by roasting sulfide ores such as zinc sulfide in air.

The sulfur dioxide and oxygen gases are then heated to a temperature of approximately 450°C and fed into a reaction vessel, where they are passed over a catalyst of vanadium(V) oxide (V_2O_5). This catalyses the reaction between sulfur dioxide and oxygen to produce sulfur trioxide (sulfur(VI) oxide, SO_3).

sulfur dioxide + oxygen $\rightleftharpoons$ sulfur trioxide

$$2SO_2(g) + O_2(g) \rightleftharpoons 2SO_3(g)$$
$$\Delta H = -197 \text{ kJ/mol}$$

This reaction is reversible and so the ideas of Le Chatelier described in section 7.5 can be used to increase the proportion of sulfur trioxide in the equilibrium mixture. The forward reaction is exothermic and so would be favoured by low temperatures. The temperature of 450°C is an optimum temperature which produces sufficient sulfur trioxide at an economical rate.

Since the reaction from left to right is also accompanied by a decrease in the number of molecules of gas, it will be favoured by a high pressure. In reality, the process is run at 200 kPa (2 atmospheres) of pressure. Under these conditions, about 96% of the sulfur dioxide and oxygen are converted into sulfur trioxide. The energy produced by this reaction is used to heat the incoming gases, thereby saving money.

The sulfur trioxide produced can be reacted with water to produce sulfuric acid.

sulfur trioxide + water $\rightarrow$ sulfuric acid

$$SO_3(g) + H_2O(l) \rightarrow H_2SO_4(l)$$

The manufacture of sulfuric acid requires several safety considerations. These are: ensuring sulfur dust explosions are prevented by wetting down the sulfur as it is handled in the process; and carefully managing and neutralising acid spray and leaks if they occur.

Key definition

The equation for the production of sulfur trioxide in the **Contact process** is $2SO_2 + O_2 \rightleftharpoons 2SO_3$

The typical conditions used in the Contact process are 450°C, 200 kPa and vanadium(V) oxide catalyst.

Burning sulfur or sulfide ores in air is the source of sulfur dioxide and air is the source of oxygen in the Contact process.

▶ Test yourself

11 Produce a flow diagram to show the different processes which occur during the production of sulfuric acid by the Contact process. Write balanced chemical equations showing the processes which occur at the different stages, giving the essential raw materials and conditions used.

12 Both the following reactions are reversible:
   (i) $X_2(g) + O_2(g) \rightleftharpoons 2XO(g)$
   (ii) $2XO(g) + O_2(g) \rightleftharpoons 2XO_2(g)$
   Suggest a reason why an increase in pressure:
   a does not favour reaction (i)
   b increases the amount of XO_2 produced in reaction (ii).

Revision checklist

After studying Chapter 7 you should be able to:
- ✓ Identify a chemical and physical change.
- ✓ Describe | and explain | the effect of changes in concentration, pressure, surface area and temperature, and the use of a catalyst on the rate of a reaction.
- ✓ Describe | and evaluate | different methods of investigating rates of a reaction.
- ✓ Interpret data and graphs obtained from rate experiments.

- ✓ Describe the collision theory in terms of the number of particles per unit volume and frequency of the collisions between the particles.
- ✓ Explain that increasing the concentration of a solution, or the pressure of a gas, increases the number of particles per unit volume, which leads to more frequent collisions between them.
- ✓ Explain that increasing the temperature leads to the particles having more kinetic energy, making them move faster and collide more frequently causing more of the collisions to be successful.

- ✓ State that a catalyst increases the rate of a reaction and is unchanged at the end of the reaction.

- ✓ Explain that a catalyst lowers the activation energy, E_a, of a reaction which means that more of the collisions are successful.

- ✓ Use the $\rightleftharpoons$ symbol to show a reversible reaction.
- ✓ Explain when a chemical equilibrium occurs.
- ✓ Describe the effects of changing the conditions for a reversible reaction for hydrated and anhydrous compounds.

- ✓ Predict and explain how changing the conditions for a reversible reaction will alter the position of equilibrium.
- ✓ State the equation for the Haber process and the sources of the nitrogen and hydrogen needed.
- ✓ State and explain the typical conditions used in the Haber process.

Exam-style questions

1 Explain the following statements.
  a A car exhaust pipe will rust much faster if the car is in constant use. [2]
  b Vegetables cook faster when they are chopped up. [1]
  c Industrial processes become more economically viable if a catalyst can be found for the reactions involved. [2]
  d In fireworks it is usual for the ingredients to be powdered. [2]
  e Tomatoes ripen faster in a greenhouse. [2]
  f The reaction between zinc and dilute hydrochloric acid is slower than the reaction between zinc and concentrated hydrochloric acid. [2]

2 A student performed two experiments to establish how effective manganese(IV) oxide was as a catalyst for the decomposition of hydrogen peroxide. The results below were obtained by carrying out these experiments with two different quantities of manganese(IV) oxide. The volume of the gas produced was recorded against time.

Time/s	0	30	60	90	120	150	180	210
Volume for 0.3 g/cm³	0	29	55	79	98	118	133	146
Volume for 0.5 g/cm³	0	45	84	118	145	162	174	182

  a Draw a diagram of the apparatus you could use to carry out these experiments. [3]
  b Plot a graph of the results. [3]
  c Is the manganese(IV) oxide acting as a catalyst in this reaction? Explain your answer. [2]
  d i At which stage does the reaction proceed most quickly? [1]
    ii How can you tell this from your graph? [1]
    iii In terms of particles, explain why the reaction is quickest at the point you have chosen in i. [2]
  e Why does the slope of the graph become less steep as the reaction proceeds? [2]
  f What volume of gas has been produced when using 0.3 g of manganese(IV) oxide after 50 s? [1]

  g How long did it take for 60 cm³ of gas to be produced when the experiment was carried out using 0.5 g of the manganese(IV) oxide? [1]
  h Write a balanced chemical equation for the decomposition of hydrogen peroxide. [1]

3 A flask containing dilute hydrochloric acid was placed on a digital balance. An excess of limestone chippings was added to this acid, a plug of cotton wool was placed in the neck of the flask and the initial mass was recorded. The mass of the apparatus was recorded every two minutes. At the end of the experiment, the loss in mass of the apparatus was calculated and the following results were obtained.

Time/min	0	2	4	6	8	10	12	14	16
Loss in mass/g	0	2.1	3.0	3.1	3.6	3.8	4.0	4.0	4.0

  a Plot the results of the experiment. [3]
  b Which of the results would appear to be incorrect? Explain your answer. [2]
  c Write a balanced chemical equation to represent the reaction taking place. [1]
  d Why did the mass of the flask and its contents decrease? [1]
  e Why was the plug of cotton wool used? [1]
  f How does the rate of reaction change during this reaction? Explain this using particle theory. [2]
  g How long did the reaction last? [1]
  h How long did it take for half of the reaction to occur? [1]

4 a What is a catalyst? [2]
  b List the properties of catalysts. [3]
  c Name the catalyst used in the following processes:
    i the Haber process [1]
    ii the hydrogenation of unsaturated fats. [1]
  d Which series of metallic elements in the Periodic Table (p. 135) do the catalysts you have named in c belong to? [1]
  e What are the conditions used in the industrial processes named in c? [2]

5 Suggest practical methods by which the rate of reaction can be investigated in each of the following cases:
  a magnesium reacting with hydrochloric acid [2]
  b nitrogen monoxide reacting with oxygen. [2]

6 Ammonia gas is made industrially by the Haber process, which involves the reaction between the gases nitrogen and hydrogen. The amount of ammonia gas produced from this reaction is affected by both the temperature and the pressure at which the process is run. The graph shows how the amount of ammonia produced from the reaction changes with both temperature and pressure. The percentage yield of ammonia indicates the percentage of the nitrogen and hydrogen gases that are actually changed into ammonia gas.

  a Write a word and balanced chemical equation for the reversible reaction between nitrogen and hydrogen to produce ammonia using the Haber process. [2]
  b What is meant by the term 'reversible reaction'? [2]
  c Use the graphs to say whether more ammonia is produced at:
    i higher or lower temperatures [1]
    ii higher or lower pressures. [1]
  d What is the percentage yield of ammonia if the conditions used to run the process are:
    i a temperature of 350°C and a pressure of 10 000 kPa? [1]
    ii a temperature of 550°C and a pressure of 35 000 kPa? [1]
  e The conditions in industry for the production of ammonia are commonly of the order of 20 000 kPa and 450°C. What is the percentage yield of ammonia using these conditions? [1]
  f Why does industry use the conditions stated in part e if it is possible to obtain a higher yield of ammonia using different conditions? [2]

Acids, bases and salts

In this chapter, you will look at acids and bases, their properties and their reactions. You will look at the use of indicators to tell whether a solution is acidic or alkaline, and you will be introduced to indicators other than universal indicator solution, which you have probably already met.

The reactions we will look at in this chapter produce ionic solids called salts. Salts are very important to our lives: for example, ammonium nitrate is the most-used fertiliser on our planet and magnesium sulfate is used in medicines. This chapter will guide you through different methods of making salts and which reactants you would use. For example, how do you make ammonium nitrate? Which reactants will you need and which of the different methods you will read about should you use?

By the end of this chapter you should be able to answer these questions. You will also become more confident in writing chemical equations and using these equations to work out the quantities of materials you need to start with and how much of the product you should make.

8.1 Acids and alkalis

All the substances shown in Figure 8.1 contain an **acid** of one sort or another. Acids are certainly all around us. What properties do these substances have which make you think that they are acids or contain acids?

▲ **Figure 8.1** What do all these foods and drinks have in common?

Some common alkaline substances are shown in Figure 8.2.

▲ **Figure 8.2** Some common alkaline substances

It would be too dangerous to taste a liquid to find out if it was acidic. Chemists use substances called **indicators** which change colour when they are added to acids or alkalis. Many indicators are dyes which have been extracted from natural sources, such as litmus.

Methyl orange, thymolphthalein and blue and red litmus are all indicators used in titrations (see p. 128). Table 8.1 shows the colours they turn in acids and alkalis.

▼ **Table 8.1** Indicators and their colours in acid and alkaline solution

Indicator	Colour in acid solution	Colour in alkaline solution
Blue litmus	Red	Blue
Methyl orange	Pink	Yellow
Thymolphthalein	Colourless	Blue
Red litmus	Red	Blue

These indicators tell chemists whether a substance is acid or alkaline (Figure 8.3). To obtain an idea of how acidic or alkaline a substance is, we use another indicator known as a universal indicator. This indicator is a mixture of many other indicators. The colour shown by this indicator can be matched against a **pH scale**. The pH scale was developed by a Scandinavian chemist called Søren Sørenson. The pH scale runs from below 0 to 14. A substance with a pH of less than 7 is an acid. One with a pH of greater than 7 is alkaline. One with a pH of 7 is said to be neither acid nor alkaline: it is neutral. Water is the most common example of a neutral substance. Figure 8.4 shows the universal indicator colour range along with everyday substances with their particular pH values.

▲ **Figure 8.3** Indicators tell you if a substance is acid or alkaline

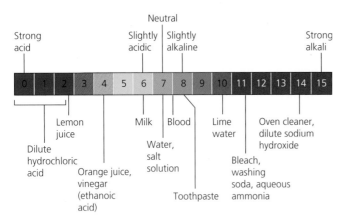

Strong acid — Slightly acidic — Neutral — Slightly alkaline — Strong alkali

Lemon juice
Dilute hydrochloric acid
Orange juice, vinegar (ethanoic acid)
Milk
Water, salt solution
Toothpaste
Blood
Lime water
Bleach, washing soda, aqueous ammonia
Oven cleaner, dilute sodium hydroxide

a   The pH scale

b   Universal indicator in solution, showing the colour range

▲ Figure 8.4

Using universal indicator allows us to determine how concentrated one acid solution is compared to another. The redder the colour, the more acidic the solution is. Solutions which are very acidic contain higher concentrations of hydrogen ions ($H^+(aq)$).

Another way in which the pH of a substance can be measured is by using a pH meter (Figure 8.5). The pH electrode is placed into the solution and a pH reading is given on the digital display.

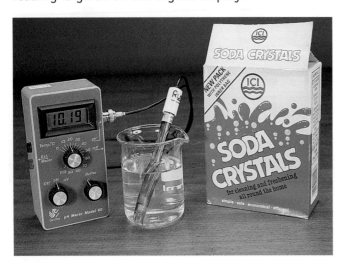

▲ Figure 8.5 A digital pH meter

The Brønsted–Lowry theory

This theory defined:

» an **acid** as an H^+ ion (or proton) donor
» a **base** as an H^+ ion (or proton) acceptor.

> **Key definitions**
>
> **Acids** are proton donors.
>
> **Bases** are proton acceptors.

The theory explains why a pure acid behaves differently from its aqueous solution, since for an acid to behave as an H^+ ion donor it must have another substance present to accept the H^+ ion. So the water, in the aqueous acid solution, is behaving as a Brønsted–Lowry base and accepting an H^+ ion. Generally:

$$HA(aq) + H_2O(l) \rightarrow H_3O^+(aq) + A^-(aq)$$

$$acid \qquad base$$

If a substance can behave both as a proton acceptor (base) and a proton donor (acid) then it is called **amphoteric**, for example, aluminium oxide, Al_2O_3 and zinc oxide, ZnO.

The relative strengths of acids and bases

The relative strength of an acid is found by comparing one acid with another. The strength of any acid depends upon how many molecules dissociate (or ionise) when the acid is dissolved in water. The relative strength of a base is found by comparing one base with another and is again dependent upon the dissociation of the base in aqueous solution.

Strong and weak acids

A typical strong acid is hydrochloric acid. It is formed by dissolving hydrogen chloride gas in water. In hydrochloric acid, the ions formed separate completely.

hydrogen chloride $\xrightarrow{\text{water}}$ hydrogen + chloride
$$\text{ions} \qquad \text{ions}$$

$$HCl(g) \qquad \xrightarrow{\text{water}} \quad H^+(aq) + Cl^-(aq)$$

> **Key definition**
>
> A **strong acid** is an acid that is completely dissociated in aqueous solution. Hydrochloric acid is a strong acid: $HCl(aq) \rightarrow H^+(aq) + Cl^-(aq)$.

For hydrochloric acid, *all* the hydrogen chloride molecules break up to form H^+ ions and Cl^- ions. Any acid that behaves in this way is termed a **strong acid**. Both sulfuric acid and nitric acid also behave in this way and are therefore also termed strong acids. All these acids have a high concentration of hydrogen ions in solution ($H^+(aq)$) and have a low pH. Their solutions are good conductors of electricity and they react quickly with metals, bases and metal carbonates.

When strong acids are neutralised by strong alkalis, the following reaction takes place between hydrogen ions and hydroxide ions.

$$H^+(aq) + OH^-(aq) \rightarrow H_2O(l)$$

A **weak acid**, such as ethanoic acid, which is found in vinegar, produces few hydrogen ions when it dissolves in water compared with a strong acid of the same concentration. It is only partially dissociated. Its solution has a higher pH than a strong acid, but still less than 7.

ethanoic acid $\underset{\text{water}}{\rightleftharpoons}$ hydrogen ions + ethanoate ions

$$CH_3COOH(aq) \underset{\text{water}}{\rightleftharpoons} \quad H^+(aq) \quad + CH_3COO^-(aq)$$

> **Key definition**
>
> A **weak acid** is an acid that is partially dissociated in aqueous solution. Ethanoic acid is a weak acid: $CH_3COOH(aq) \rightleftharpoons CH_3COO^-(aq) + H^+(aq)$.

The $\rightleftharpoons$ sign means that the reaction is reversible. This means that if the ethanoic acid molecule breaks down to give hydrogen ions and ethanoate

ions then they will react together to re-form the ethanoic acid molecule. The fact that fewer ethanoic acid molecules dissociate compared with a strong acid, and that the reaction is reversible, means that few hydrogen ions are present in the solution. Other examples of weak acids are citric acid, found in oranges and lemons; carbonic acid, found in soft drinks; sulfurous acid (acid rain) (Figure 8.6) and ascorbic acid (vitamin C).

▲ **Figure 8.6** Sulfurous acid is found in acid rain. It is a weak acid and is oxidised to sulfuric acid (a strong acid). Acid rain damages the environment quite badly

Solutions of weak acids are poorer conductors of electricity and have slower reactions with metals, bases and metal carbonates.

All acids when in aqueous solution produce hydrogen ions, $H^+(aq)$. To say an acid is a strong acid does not mean it is concentrated. The *strength* of an acid tells you how easily it dissociates (ionises) to produce hydrogen ions. The *concentration* of an acid indicates the proportions of water and acid present in aqueous solution. It is important to emphasise that a strong acid is still a strong acid even when it is in dilute solution and a weak acid is still a weak acid even if it is concentrated.

Neutralisation reactions

A common situation involving neutralisation of an acid is when you suffer from indigestion. This is caused by acid in your stomach irritating the stomach lining or throat. Normally it is treated by taking an indigestion remedy containing a substance which will react with and neutralise the acid.

In the laboratory, if you wish to neutralise a common acid such as hydrochloric acid you can use an alkali such as sodium hydroxide. If the pH of the acid is measured when some sodium hydroxide solution is added to it, the pH increases. If equal volumes of the same concentration of hydrochloric acid and sodium hydroxide are added to one another, the resulting solution is found to have a pH of 7. The acid has undergone a **neutralisation reaction** and a neutral solution has been formed.

$$\begin{array}{ccccccc} \text{hydrochloric} & + & \text{sodium} & \rightarrow & \text{sodium} & + & \text{water} \\ \text{acid} & & \text{hydroxide} & & \text{chloride} & & \end{array}$$

$$HCl(aq) + NaOH(aq) \rightarrow NaCl(aq) + H_2O(l)$$

> **Key definition**
>
> A **neutralisation reaction** occurs between an acid and a base to produce water; $H^+(aq) + OH^-(aq) \rightarrow H_2O(l)$.

As we have shown, when both hydrochloric acid and sodium hydroxide dissolve in water, the ions separate completely. We may therefore write:

$$H^+(aq)Cl^-(aq) + Na^+(aq)OH^-(aq) \rightarrow Na^+(aq) + Cl^-(aq) + H_2O(l)$$

You will notice that certain ions are unchanged on either side of the equation. They are called spectator ions and are usually taken out of the equation. The equation now becomes:

$$H^+(aq) + OH^-(aq) \rightarrow H_2O(l)$$

This type of equation is known as an **ionic equation**. The reaction between any acid and any alkali in aqueous solution produces water and can be summarised by this ionic equation. It shows the ion which causes acidity ($H^+(aq)$) reacting with the ion which causes alkalinity ($OH^-(aq)$) to produce neutral water ($H_2O(l)$). Aqueous solutions of acids contain H^+ ions and aqueous solutions of alkalis contain OH^- ions.

> ▶ **Test yourself**
>
> 1 Complete the following balanced chemical equations.
>   a $CH_3COOH + NaOH \rightarrow$
>   b $H_2SO_4 + KOH \rightarrow$
>   c $NH_3 + HBr \rightarrow$
>   In each case, name the acid and the base.
>
> ---
>
> 2 Explain the terms 'concentrated' and 'strength' as applied to acids.
>
> ---
>
> 3 Explain what part water plays in the acidity of a solution.
> 4 a Write a chemical equation to represent the neutralisation of sulfuric acid by sodium hydroxide.
>   b Reduce this to an ionic equation.
>   c Account for any difference you see between the ionic equation you have written and the one shown on the left for the reaction of hydrochloric acid and sodium hydroxide.

8.2 Formation of salts

In the example on this page, sodium chloride was produced as part of the neutralisation reaction. Compounds formed in this way are known as salts. A salt is a compound that has been formed when all the hydrogen ions of an acid have been replaced by metal ions or by the ammonium ion (NH_4^+).

Salts can be classified as those which are soluble in water or those which are insoluble in water. The general solubility rules for salts are:

» all sodium, potassium and ammonium salts are soluble
» all nitrates are soluble
» all chlorides are soluble, except lead and silver
» all sulfates are soluble, except barium, calcium and lead
» all carbonates and hydroxides are insoluble, except sodium, potassium and ammonium.

If the acid being neutralised is hydrochloric acid, salts called chlorides are formed. Other types of salts can be formed with other acids. A summary of the different types of salt along with the acid they are formed from is shown in Table 8.2.

▼ **Table 8.2** Types of salt and the acids they are formed from

Acid	Type of salt	Example
Carbonic acid	Carbonates	Sodium carbonate (Na_2CO_3)
Ethanoic acid	Ethanoates	Sodium ethanoate (CH_3COONa)
Hydrochloric acid	Chlorides	Potassium chloride (KCl)
Nitric acid	Nitrates	Potassium nitrate (KNO_3)
Sulfuric acid	Sulfates	Sodium sulfate (Na_2SO_4)

Types of oxides

There are three different types of oxides. Non-metal oxides, such as sulfur dioxide, SO_2, and carbon dioxide are acidic. In aqueous solution they produce aqueous hydrogen ions, $H^+(aq)$. Metal oxides, however, are basic oxides. If these oxides are soluble they will dissolve in water to produce aqueous hydroxide ions, $OH^-(aq)$. For example, copper(II) oxide, CuO and calcium oxide, CaO.

The third type of oxides, and the rarest, are those described as **amphoteric oxides**. These oxides react with both acids and alkalis to produce salts. Examples of amphoteric oxides are zinc oxide, ZnO, and aluminium oxide, Al_2O_3.

8.3 Methods of preparing soluble salts

There are four general methods of preparing soluble salts. These involve the reaction of an acid with:

» excess metal
» excess insoluble carbonate
» excess insoluble base
» an alkali by titration.

Acid + metal

acid + metal → salt + hydrogen

This method can only be used with the less reactive metals such as aluminium. It would be very dangerous to use a reactive metal, such as sodium, in this type of reaction. The metals usually used in this method of salt preparation are the **MAZIT** metals, that is, **m**agnesium, **a**luminium, **z**inc, **i**ron and **t**in. A typical experimental method is given below.

Excess magnesium ribbon is added to dilute hydrochloric acid. By using an excess of magnesium ribbon we are making sure that all of the acid has reacted and some magnesium is left at the end of the reaction. During this addition an effervescence is observed due to the production of hydrogen gas. In this reaction the hydrogen ions from the hydrochloric acid gain electrons from the metal atoms as the reaction proceeds.

hydrogen ions + electrons → hydrogen gas
(from metal)

$$2H^+ \quad + \quad 2e^- \quad \rightarrow \quad H_2(g)$$

magnesium + hydrochloric → magnesium + hydrogen
acid chloride

$$Mg(s) \quad + 2HCl(aq) \rightarrow MgCl_2(aq) + H_2(g)$$

▲ **Figure 8.7** The excess magnesium is filtered in this way

The excess magnesium is removed by filtration (Figure 8.7).

The magnesium chloride solution is evaporated slowly to form a **saturated** solution of the salt (Figure 8.8).

The hot concentrated magnesium chloride solution produced is tested by dipping a cold glass rod into it. If salt crystals form at the end of the rod the solution is ready to crystallise and is left to cool. Any crystals produced on cooling are filtered and dried between clean tissues.

▲ **Figure 8.8** The solution of magnesium nitrate is concentrated by slow evaporation

Acid + carbonate

acid + carbonate → salt + water + carbon dioxide

This method can be used with any metal carbonate and any acid, providing the salt produced is soluble. The typical experimental procedure is similar to that carried out for an acid and a metal. For example, copper(II) carbonate would be added in excess to dilute nitric acid. Effervescence would be observed due to the production of carbon dioxide.

| copper(II) carbonate | + | nitric acid | → | copper(II) nitrate | + carbon dioxide | + water |

$$CuCO_3(s) + 2HNO_3(aq) → Cu(NO_3)_2(aq) + CO_2(g) + H_2O(l)$$

Metal carbonates contain carbonate ions, CO_3^{2-}. In this reaction the carbonate ions react with the hydrogen ions in the acid.

carbonate + hydrogen → carbon + water
ions   ions   dioxide

$$CO_3^{2-}(aq) + 2H^+(aq) → CO_2(g) + H_2O(l)$$

When excess carbonate has been added, you will see it collect at the bottom of the beaker and the effervescence will stop.

The excess copper(II) carbonate can be filtered off and the copper(II) nitrate solution is evaporated slowly to form a saturated solution of the salt.

The hot concentrated copper(II) nitrate solution produced is tested by dipping a cold glass rod into it. If salt crystals form at the end of the rod the solution is ready to crystallise and is left to cool. Any crystals produced on cooling are filtered and dried between clean tissues.

Acid + alkali (soluble base)

acid + alkali → salt + water

Titration is generally used for preparing the salts of very reactive metals, such as potassium or sodium. It would be too dangerous to add the metal directly to the acid. In this case, we solve the problem indirectly and use an alkali which contains the reactive metal whose salt we wish to prepare.

Most metal oxides and hydroxides are insoluble bases: they are said to be basic. A few metal oxides and hydroxides that do dissolve in water to produce OH^-(aq) ions are known as **alkalis**, or **soluble bases**. If the metal oxide or hydroxide does not dissolve in water, it is known as an **insoluble base**.

> **Key definitions**
>
> **Alkalis** are soluble bases. Aqueous solutions of alkalis contain OH^- ions.

A **base** is a substance which neutralises an acid, producing a salt and water as the only products. If the base is soluble, the term 'alkali' can be used, but there are several bases which are insoluble. It is also a substance which accepts a hydrogen ion (see p. 118). In general, most metal oxides and hydroxides (as well as ammonia solution) are bases. Some examples of soluble and insoluble bases are shown in Table 8.3. Salts can be formed by this method only if the base is soluble.

▼ **Table 8.3** Examples of soluble and insoluble bases

Soluble bases (alkalis)	Insoluble bases
Sodium hydroxide (NaOH)	Iron(III) oxide (Fe_2O_3)
Potassium hydroxide (KOH)	Copper(II) oxide (CuO)
Calcium hydroxide (Ca(OH)$_2$)	Lead(II) oxide (PbO)
Ammonia solution (NH$_3$(aq))	Magnesium oxide (MgO)

In this neutralisation reaction both reactants are in solution, so a special technique called titration is required. Acid is slowly and carefully added to a measured volume of alkali using a burette (Figure 8.9) until the indicator, for example thymolphthalein, changes colour.

An indicator is used to show when the alkali has been neutralised completely by the acid. This is called the end-point. Once you know where the end-point is, you can add the same volume of acid to the measured volume of alkali but this time without the indicator.

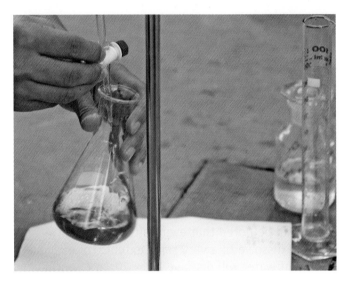

▲ **Figure 8.9** The acid is added to the alkali until the indicator just changes colour

For example, the salt sodium chloride can be made by reacting sodium hydroxide with hydrochloric acid using this method.

$$\begin{array}{cccc} \text{hydrochloric} + & \text{sodium} & \rightarrow & \text{sodium} & + \text{water} \\ \text{acid} & \text{hydroxide} & & \text{chloride} \end{array}$$

$$HCl(aq) + NaOH(aq) \rightarrow NaCl(aq) + H_2O(l)$$

As previously discussed on p. 120, this reaction can best be described by the ionic equation:

$$H^+(aq) + OH^-(aq) \rightarrow H_2O(l)$$

The sodium chloride solution which is produced can then be evaporated slowly to obtain the salt.

The hot concentrated sodium chloride solution produced is tested by dipping a cold glass rod into it. If salt crystals form at the end of the rod, the solution is ready to crystallise and is left to cool. Any crystals produced on cooling are filtered and dried between clean tissues.

Acid + insoluble base

$$\text{acid} + \text{base} \rightarrow \text{salt} + \text{water}$$

This method can be used to prepare a salt of an unreactive metal, such as lead or copper. In these cases it is not possible to use a direct reaction of the metal with an acid so the acid is neutralised using the particular metal oxide (Figure 8.10).

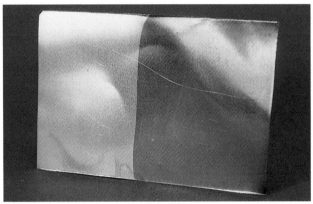

▲ **Figure 8.10** Citric acid has been used on the right-hand side of this piece of copper metal to remove the oxide coating on its surface. The acid reacts with the oxide coating to form water, leaving shiny copper metal

The method is generally the same as that for a metal carbonate and an acid, though some warming of the reactants may be necessary. An example of such a reaction is the neutralisation of sulfuric acid by copper(II) oxide to produce copper(II) sulfate (Figure 8.11).

▲ **Figure 8.11** After slow evaporation to concentrate the solution, the solution is left to crystallise. Crystals of copper(II) sulfate are produced

$$\begin{array}{cccc} \text{sulfuric} & + \text{copper(II)} \rightarrow & \text{copper(II)} & + \text{water} \\ \text{acid} & \text{oxide} & \text{sulfate} \end{array}$$

$$H_2SO_4(aq) + CuO(s) \rightarrow CuSO_4(aq) + H_2O(l)$$

Metal oxides contain the oxide ion, O^{2-}. The ionic equation for this reaction is therefore:

$$2H^+(aq) + O^{2-}(s) \rightarrow H_2O(l)$$

or

$$CuO(s) + 2H^+(aq) \rightarrow Cu^{2+}(aq) + H_2O(l)$$

Reaction of a base with ammonium salts

Small quantities of ammonia gas, NH_3, can be produced by heating any ammonium salt, such as ammonium chloride, with a base, such as calcium hydroxide. This reaction can be used to make ammonia in a laboratory.

calcium + ammonium → calcium + water + ammonia
hydroxide   chloride   chloride

$$Ca(OH)_2(s) + 2NH_4Cl \rightarrow CaCl_2 + H_2O + 2NH_3$$

The ammonia produced can be detected as being formed by its pungent odour and by turning damp red litmus blue.

8.4 Preparing insoluble salts

The methods described above can be used to make a salt that is soluble in water. If a salt that is insoluble in water needs to be prepared, a different technique is needed.

An insoluble salt, such as barium sulfate, can be made by precipitation. In this case, solutions of the two chosen soluble salts are mixed (Figure 8.12). To produce barium sulfate, barium chloride and sodium sulfate can be used. The barium sulfate precipitate can be filtered off, washed with distilled water and dried. The reaction that has occurred is:

barium + sodium → barium + sodium
chloride   sulfate   sulfate   chloride

$$BaCl_2(aq) + Na_2SO_4(aq) \rightarrow BaSO_4(s) + 2NaCl(aq)$$

The ionic equation for this reaction is:

$$Ba^{2+}(aq) + SO_4^{2-}(aq) \rightarrow BaSO_4(s)$$

▲ **Figure 8.12** When barium chloride solution is added to sodium sulfate, a white precipitate of barium sulfate forms

This method may be summarised as follows:

soluble + soluble → insoluble + soluble
salt   salt   salt   salt

$$(AX) + (BY) \rightarrow (BX) + (AY)$$

It should be noted that even salts like barium sulfate dissolve to a very small extent. For example, 1 dm³ of water will dissolve 2.2×10^{-3} g of barium sulfate at 25°C. This substance and substances like it are said to be sparingly soluble.

8.5 Testing for different salts

Sometimes we want to analyse a salt and find out what is in it. There are simple chemical tests which allow us to identify the anion part of the salt.

Testing for a sulfate (SO_4^{2-})

You have seen that barium sulfate is an insoluble salt (p. 120). Therefore, if you take a solution of a suspected sulfate and add it to a solution of a soluble barium salt (such as barium chloride) then a white precipitate of barium sulfate will be produced.

barium ion + sulfate ion → barium sulfate

$$Ba^{2+}(aq) + SO_4^{2-}(aq) \rightarrow BaSO_4(s)$$

A few drops of dilute hydrochloric acid are also added to this mixture. If the precipitate does not dissolve, then it is barium sulfate and the unknown salt was in fact a sulfate. If the precipitate does dissolve, then the unknown salt may have been a sulfite (containing the SO_3^{2-} ion).

Testing for a chloride (Cl^-), a bromide (Br^-) or an iodide (I^-)

Earlier in this chapter you saw that silver chloride is an insoluble salt (p. 120). Therefore, if you take a solution of a suspected chloride and add to it a small volume of dilute nitric acid, to make an aqueous acidic solution, followed by a small amount of a solution of a soluble silver salt (such as silver nitrate), a white precipitate of silver chloride will be produced.

chloride ion + silver ion → silver chloride

$$Cl^-(aq) + Ag^+(aq) \rightarrow AgCl(s)$$

If left to stand, the precipitate goes grey (Figure 8.13).

▲ **Figure 8.13** If left to stand, the white precipitate of silver chloride goes grey. This photochemical change plays an essential part in black and white photography

In a similar way, a bromide and an iodide will react to produce either a cream precipitate of silver bromide (AgBr) or a yellow precipitate of silver iodide (AgI) (Figure 8.14).

▲ **Figure 8.14** AgCl, a white precipitate, AgBr, a cream precipitate, and AgI, a yellow precipitate

An alternative test for iodide ions is the addition of lead nitrate solution to the iodide which results in a bright yellow precipitate of lead iodide, PbI_2.

Testing for a carbonate

If a small amount of an acid is added to some of the suspected carbonate (either solid or in solution) then effervescence occurs. If it is a carbonate then carbon dioxide gas is produced, which will turn limewater 'milky' (a cloudy white precipitate of calcium carbonate forms, see Chapter 14, p. 236).

carbonate + hydrogen → carbon + water
ions   ions   dioxide

$$CO_3^{2-}(aq) + 2H^+(aq) \rightarrow CO_2(g) + H_2O(l)$$

Testing for a nitrate

By adding aqueous sodium hydroxide and then aluminium foil and warming gently, nitrates are reduced to ammonia. The ammonia can be identified using damp indicator paper, which turns blue.

$$3NO_3^-(aq) + 8Al(s) + 5OH^-(aq) + 18H_2O(l)$$
$$\downarrow$$
$$3NH_3(g) + 8[Al(OH)_4]^-(aq)$$

In the reaction the nitrate ion is reduced, as oxygen is removed from the nitrogen atom, and it gains hydrogen to form ammonia, NH_3. The gain of hydrogen is also a definition of reduction.

Test yourself

5 Complete the word equations and write balanced chemical equations for the following soluble salt preparations.
   a magnesium + sulfuric acid →
   b calcium carbonate + hydrochloric acid →
   c zinc oxide + hydrochloric acid →
   d potassium hydroxide + nitric acid →
   Also write ionic equations for each of the reactions.
6 Lead carbonate and lead iodide are insoluble. Which two soluble salts could you use in the preparation of each substance? Write:
   a a word equation
   b a symbol equation

   c an ionic equation

   to represent the reactions taking place.
7 An analytical chemist working for an environmental health organisation has been given a sample of water which is thought to have been contaminated by a sulfate, a carbonate and a chloride.
   a Describe how she could confirm the presence of these three types of salt by simple chemical tests.

   b Write ionic equations to help you explain what is happening during the testing process.

8.6 Water of crystallisation

Some salts, such as sodium chloride, copper carbonate and sodium nitrate, crystallise in their **anhydrous** forms (without water). However, many salts produce **hydrates** when they crystallise from solution. A hydrate is a salt which incorporates water into its crystal structure. This water is referred to as **water of crystallisation**. The shape of the crystal hydrate is very much dependent on the presence of water of crystallisation. Some examples of crystal hydrates are given in Table 8.4 and shown in Figure 8.15.

> **Key definitions**
>
> A hydrated substance, or **hydrate**, is one that is chemically combined with water.
>
> An **anhydrous** substance is one containing no water.

▼ **Table 8.4** Examples of crystal hydrates

Salt hydrate	Formula
Cobalt(II) chloride hexahydrate	$CoCl_2.6H_2O$
Copper(II) sulfate pentahydrate	$CuSO_4.5H_2O$
Iron(II) sulfate heptahydrate	$FeSO_4.7H_2O$
Magnesium sulfate heptahydrate	$MgSO_4.7H_2O$

▲ **Figure 8.15** Hydrate crystals (left to right): cobalt nitrate, calcium nitrate and nickel sulfate (top) and manganese sulfate, copper sulfate and chromium potassium sulfate (bottom)

When many hydrates are heated, the water of crystallisation is driven away. For example, if crystals of copper(II) sulfate hydrate (blue) are heated strongly, they lose their water of crystallisation. Anhydrous copper(II) sulfate remains as a white powder:

copper(II) sulfate → anhydrous copper(II) + water
   pentahydrate   sulfate

$$CuSO_4.5H_2O(s) \rightarrow CuSO_4(s) + 5H_2O(g)$$

When water is added to anhydrous copper(II) sulfate, the reverse process occurs. It turns blue and the pentahydrate is produced (Figure 8.16). This is an extremely exothermic process.

$$CuSO_4(s) + 5H_2O(l) \rightarrow CuSO_4.5H_2O(s)$$

> **Key definition**
>
> **Water of crystallisation** is the water molecules present in crystals, e.g. $CuSO_4.5H_2O$ and $CoCl_2.6H_2O$.

▲ **Figure 8.16** Anhydrous copper(II) sulfate is a white powder which turns blue when water is added to it

Because the colour change only takes place in the presence of water, the reaction is used to test for the presence of water.

These processes give a simple example of a reversible reaction:

$$CuSO_4(s) + 5H_2O(l) \rightleftharpoons CuSO_4.5H_2O(s)$$

 Going further

Calculation of water of crystallisation
Sometimes it is necessary to work out the percentage, by mass, of water of crystallisation in a hydrated salt.

 Worked example

Calculate the percentage by mass of water in the salt hydrate $MgSO_4.7H_2O$. (A_r: H = 1; O = 16; Mg = 24; S = 32)

M_r for $MgSO_4.7H_2O$

$$= 24 + 32 + (4 \times 16) + (7 \times 18) = 246$$

The mass of water as a fraction of the total mass of hydrate

$$= \frac{126}{246}$$

The percentage of water present

$$= \frac{126}{246} \times 100$$
$$= 51.2\%$$

> **Test yourself**
>
> 8 Calculate the percentage by mass of water in the following salt hydrates:
>   a $CuSO_4.5H_2O$ b $Na_2CO_3.10H_2O$
>   c $Na_2S_2O_3.5H_2O$.
>   (A_r: H = 1; O = 16; Na = 23; S = 32; Cu = 63.5)

 Practical skills

Preparation of potassium nitrate
For safe experiments/demonstrations which are related to this chapter, please refer to the *Cambridge IGCSE Chemistry Practical Skills Workbook*, which is also part of this series.

Safety
- Eye protection must be worn.
- Take care using indicator solution as it can stain skin and clothing.

The apparatus below can be used to prepare potassium nitrate crystals.

1 Clamp a burette vertically in a clamp stand and fill it up with some dilute nitric acid solution. Allow some of the acid to run through the jet of the burette. Record the initial burette reading.

2 Using a pipette and safety filler, place 25 cm^3 of the potassium hydroxide solution into a 250 cm^3 conical flask.

3 Put 4 drops of the indicator thymolphthalein into the potassium hydroxide in the flask. The solution will turn blue.

4 Slowly add the dilute nitric acid from the burette, swirling the flask all the time, until the blue colour disappears. Record the final burette reading and hence the amount of dilute nitric acid you needed to add to neutralise the potassium hydroxide.

5 Pour some of the solution formed in step 4 from the flask into an evaporating basin and heat until half of the solution has evaporated.

6 Set aside and allow the crystals to form slowly.

Now answer the following questions:

1 Why is it important to clamp the burette vertically?

2 Why is it important to fill up the jet of the burette before taking the initial burette reading?

3 Why is it important to use a safety filler when using the pipette?

4 How do you take the initial volume reading on the burette?

5 Why is the swirling of the flask important?

6 How could you tell that the solution has become saturated as it is evaporated?

7 In step 5, why is it important to not evaporate all the water away?

8 How can the crystals be dried?

Titration

On p. 122 you saw that it was possible to prepare a soluble salt by reacting an acid with a soluble base (alkali). The method used was that of titration. Titration can also be used to find the concentration of the alkali used. In the laboratory, the titration of hydrochloric acid with sodium hydroxide is carried out in the following way.

1 25 cm^3 of sodium hydroxide solution is pipetted into a conical flask to which a few drops of thymolphthalein indicator have been added (Figure 8.17). Thymolphthalein is blue in alkaline conditions but colourless in acid.

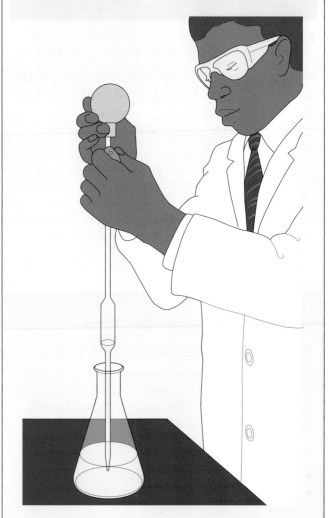

▲ **Figure 8.17** Exactly 25.0 cm^3 of sodium hydroxide solution is pipetted into a conical flask

2 A 0.10 mol/dm³ solution of hydrochloric acid is placed in the burette using a filter funnel until it is filled up exactly to the zero mark (Figure 8.18).

Dilute hydrochloric acid

Burette

Note: Remove filter funnel before titrating

▲ **Figure 8.18** The burette is filled up to the zero mark with a 0.10 mol/dm³ solution of hydrochloric acid

3 The filter funnel is now removed.
4 The hydrochloric acid is added to the sodium hydroxide solution in small quantities – usually no more than 0.5 cm³ at a time (Figure 8.19). The contents of the flask must be swirled after each addition of acid for thorough mixing.
5 The acid is added until the alkali has been neutralised completely. This is shown by the blue colour of the indicator just disappearing.
6 The final reading on the burette at the end-point is recorded and further titrations carried out until consistent results are obtained (within 0.1 cm³ of each other). Some sample data are shown on the right.

▲ **Figure 8.19** The titration is carried out accurately

Volume of sodium hydroxide solution = 25.0 cm³

Average volume of 0.10 mol/dm³ solution of hydrochloric acid added = 21.0 cm³

The neutralisation reaction which has taken place is:

hydrochloric + sodium → sodium + water
acid   hydroxide   chloride

$$HCl(aq) + NaOH(aq) \rightarrow NaCl(aq) + H_2O(l)$$

From this equation, it can be seen that 1 mole of hydrochloric acid neutralises 1 mole of sodium hydroxide.

Now you can work out the number of moles of the acid using the formula given in Chapter 4, p. 59.

$$moles = \frac{volume}{1000} \times concentration$$

$$= 21.0 \times \frac{0.10}{10000}$$

$$= 21.0 \times 10^{-3}$$

number of moles of hydrochloric acid = number of moles of sodium hydroxide

Therefore, the number of moles of sodium hydroxide = 2.1×10^{-3}

2.1×10^{-3} moles of sodium hydroxide is present in 25.0 cm³ of solution.

Therefore, in 1 cm³ of sodium hydroxide solution we have:

$$\frac{2.1 \times 10^{-3}}{25.0} \text{ moles}$$

Therefore, in 1 dm³ of sodium hydroxide solution we have:

$$\frac{2.1 \times 10^{-3}}{25.0} \times 1000 = 0.084 \text{ mole}$$

The concentration of sodium hydroxide solution is 0.084 mol/dm³.

You can simplify the calculation by substituting in the following mathematical equation:

$$\frac{M_1 V_1}{M_{acid}} = \frac{M_2 V_2}{M_{alkali}}$$

where:

M_1 = concentration of the acid used

V_1 = volume of acid used (cm³)

M_{acid} = number of moles of acid shown in the chemical equation

M_2 = concentration of the alkali used

V_2 = volume of the alkali used (cm³)

M_{alkali} = number of moles of alkali shown in the chemical equation.

In the example:

M_1 = 0.10 mol/dm³

V_1 = 21.0 cm³

M_{acid} = 1 mole

M_2 = unknown

V_2 = 25.0 cm³

M_{alkali} = 1 mole

Substituting in the equation:

$$\frac{0.10 \times 21.0}{1} = \frac{M_2 \times 25.0}{1}$$

Rearranging:

$$M_2 = \frac{0.10 \times 21.0 \times 1}{1 \times 25.0}$$

$$M_2 = 0.084$$

The concentration of the sodium hydroxide solution is 0.084 mol/dm³.

Another example of a titration calculation could involve a neutralisation reaction in which the ratio of the number of moles of acid to alkali is not 1 : 1. The example below shows how such a calculation could be carried out.

? Worked example

In a titration to find the concentration of a solution of sulfuric acid, 25 cm³ of it was just neutralised by 20.15 cm³ of a 0.2 mol/dm³ solution of sodium hydroxide. What is the concentration of the sulfuric acid used?

First, write out the balanced chemical equation for the reaction taking place.

sulfuric acid + sodium hydroxide → sodium sulfate + water

$$H_2SO_4 + 2NaOH \longrightarrow Na_2SO_4 + 2H_2O$$

From this balanced equation, it can be seen that 1 mole of sulfuric acid reacts with 2 moles of sodium hydroxide.

Therefore, the number of moles of sodium hydroxide used:

$$= 20.15 \times \frac{0.2}{1000} = 4.03 \times 10^{-3}$$

The number of moles of sulfuric acid which will react with 4.03×10^{-3} moles of sodium hydroxide:

$$= 4.03 \times 10^{-3} \times \frac{1}{2} = 2.015 \times 10^{-3}$$

This is the number of moles of sulfuric acid present in 25 cm³ of the solution, so the concentration of the sulfuric acid is:

$$2.015 \times 10^{-3} \times \frac{1000}{25} = 0.081 \text{ mol/dm}^3$$

Test yourself

9 24.2 cm³ of a solution containing 0.20 mol/dm³ of hydrochloric acid just neutralised 25.0 cm³ of a potassium hydroxide solution. What is the concentration of this potassium hydroxide solution?

10 22.4 cm³ of a solution containing 0.10 mol/dm³ of sulfuric acid just neutralised 25.0 cm³ of a sodium hydroxide solution. What is the concentration of this sodium hydroxide solution?

Revision checklist

After studying Chapter 8 you should be able to:
✓ Describe the chemical properties of acids and bases.
✓ State that an alkali is a soluble base.
✓ Explain the difference between a concentrated and dilute acid.
✓ Describe the use of indicators to determine the presence of an acid or alkali and to determine their strengths.

✓ Explain the difference between a strong and weak acid.

✓ Explain what a neutralisation reaction is.
✓ Classify oxides as being acidic, basic or amphoteric .

✓ Describe the four different methods of making soluble salts.
✓ State the rules of solubility.

✓ Explain how an insoluble salt is made.

✓ Define the terms 'hydrated' and 'anhydrous'.

Exam-style questions

1 Explain, with the aid of examples, what you understand by the following terms:

a strong acid [2]
b weak acid [2]
c strong alkali [2]
d weak base. [2]

2 a Copy out and complete the table, which covers the different methods of preparing salts.

Method of preparation	Name of salt prepared	Two substances used in the preparation
Acid + alkali	Potassium sulfate	. . . . and . . . . [2]
Acid + metal	. . . . [1]	. . . . and dilute hydrochloric acid [1]
Acid + insoluble base	Magnesium sulfate	. . . . and . . . . [2]
Acid + carbonate	Copper . . . . . [1]	. . . . and . . . . [2]
Precipitation	Lead iodide	. . . . and . . . . [2]

b Give word and balanced chemical equations for each reaction shown in your table. Also write ionic equations where appropriate. [15]

3 Consider the following scheme.

a Give the names and formulae of substances **A** to **E**. [5]
b Describe a test which could be used to identify the presence of water. [1]

c Give the indicator suitable for the initial reaction between the hydroxides and the dilute acids shown. [1]
d Give balanced chemical equations for the reactions taking place in the scheme. [3]
e Give an ionic equation for the production of the white precipitate **D**. [1]

4 In a titration involving 24.0 cm³ potassium hydroxide solution against a solution of concentration 1 mol/dm³ of sulfuric acid, 28.0 cm³ of the acid was found to just neutralise the alkali completely.

a Give a word and balanced chemical equation for the reaction. [2]
b Identify a suitable indicator for the titration and state the colour change you would observe. [1]
c Calculate the concentration of the alkali in mol/dm³. [3]
d Describe a chemical test which you could use to identify the type of salt produced during the reaction. [2]

5 For each of the following, explain the term given **and** give an example:

a neutralisation [2]
b titration [3]
c soluble salt [2]
d insoluble salt. [2]

6 Copper(II) sulfate crystals exist as the *pentahydrate*, $CuSO_4.5H_2O$. It is a salt *hydrate*. If it is heated quite strongly, the *water of crystallisation* is driven off and the *anhydrous* salt remains.

a Explain the meaning of the terms shown in italics. [4]
b Describe the experiment you would carry out to collect a sample of the water given off when the salt hydrate was heated strongly. Your description should include a diagram of the apparatus used and a chemical equation to represent the process taking place. [4]
c Describe a chemical test you could carry out to show that the colourless liquid given off was water. [2]

d Describe one other test you could carry out to show that the colourless liquid obtained in this experiment was pure water. [1]

e Sometimes it is necessary to work out the percentage by mass of water of crystallisation as well as the number of moles of water present in a hydrated crystal.

  i Use the information given to calculate the percentage, by mass, of water of crystallisation in a sample of hydrated magnesium sulfate. [3]

    Mass of crucible = 14.20 g

    Mass of crucible + hydrated $MgSO_4$ = 16.66 g

    Mass after heating = 15.40 g

  ii Calculate the number of moles of water of crystallisation driven off during the experiment as well as the number of moles of anhydrous salt remaining. [2]
    (A_r: H = 1; O = 16; Mg = 24; S = 32)

  iii Using the information you have obtained in **ii**, write down, in the form $MgSO_4.xH_2O$, the formula of hydrated magnesium sulfate. [1]

9 The Periodic Table

FOCUS POINTS
★ How can the Periodic Table help us to understand the way chemical elements react?
★ What are the trends in properties and reactivity for different groups in the Periodic Table?

In this chapter, we will look at the way in which all the chemical elements have been organised into a table that can help to predict their properties and their chemical reactions. Chemists began to try to organise the elements around 150 years ago but found it very difficult and, as new elements were discovered, they had to revise their attempts. Eventually the format of the Periodic Table which we see today was produced and, as new elements were discovered, they fitted in perfectly.

The aim of this chapter is to help you to use the Periodic Table to understand chemistry more easily. For example, if you know some of the properties of sodium, the Periodic Table can help you to predict the properties of francium.

By the end of this chapter you will have learned the trends in properties of four groups of elements and you should understand how useful the Periodic Table is to your study of chemistry.

9.1 Development of the Periodic Table

The Periodic Table is a vital tool used by chemists to predict the way in which elements react during chemical reactions. It is a method of categorising elements according to their properties. Scientists started to look for a way in which to categorise the known elements around 150 years ago.

The Periodic Table was devised in 1869 by the Russian Dmitri Mendeleev, who was the Professor of Chemistry at St Petersburg University (Figure 9.1). His Periodic Table was based on the chemical and physical properties of the 63 elements that had been discovered at that time.

Other scientists had attempted to categorise the known elements from the early 19th century but Mendeleev's classification proved to be the most successful.

 Going further

Mendeleev arranged all the 63 known elements in order of increasing atomic weight but in such a way that elements with similar properties were in the same vertical column. He called the vertical columns **groups** and the horizontal rows **periods** (Figure 9.2). If necessary, he left gaps in the table.

As a scientific idea, Mendeleev's Periodic Table was tested by making predictions about elements that were unknown at that time but could possibly fill the gaps. Three of these gaps are shown by the symbols * and † in Figure 9.2. As new elements were discovered, they were

found to fit easily into the classification. For example, Mendeleev predicted the properties of the missing element 'eka-silicon' (†). He predicted the colour, density and melting point as well as its atomic weight.

In 1886 the element we now know as germanium was discovered in Germany by Clemens Winkler; its properties were almost exactly those Mendeleev had predicted. In all, Mendeleev predicted the atomic weight of ten new elements, of which seven were eventually discovered – the other three, atomic weights 45, 146 and 175, do not exist!

▲ **Figure 9.1** Dmitri Mendeleev (1834–1907)

Period	Group							
	1	2	3	4	5	6	7	8
1	H							
2	Li	Be	B	C	N	O	F	
3	Na	Mg	Al	Si	P	S	Cl	
4	K	Ca	*	Ti	V	Cr	Mn	Fe Co Ni
	Cu	Zn	*	†	As	Se	Br	

▲ **Figure 9.2** Mendeleev's Periodic Table. He left gaps for undiscovered elements

Key

■ Reactive metals ■ Metalloids
□ Transition metals □ Non-metals
■ Poor metals ■ Noble gases

▲ **Figure 9.3** The modern Periodic Table

The success of Mendeleev's predictions showed that his ideas were probably correct. His Periodic Table was quickly accepted by scientists as an important summary of the properties of the elements.

Mendeleev's Periodic Table has been modified in the light of work carried out by Ernest Rutherford and Henry Moseley. Discoveries about sub-atomic particles led them to realise that the elements should be arranged by proton number. In the modern Periodic Table, the 118 known elements are arranged in order of increasing proton number (Figure 9.3).

Those elements with similar chemical properties are found in the same columns or **groups**. There are eight groups of elements. The first column is called Group I, the second Group II, and so on up to Group VII. The final column in the Periodic Table is called Group 0 (or Group VIII). Some of the groups have been given names.

Group I: The **alkali metals**
Group II: The **alkaline earth metals**
Group VII: The **halogens**
Group 0:   Inert gases or **noble gases**

▲ **Figure 9.4** Transition elements have a wide range of uses, both as elements and as alloys

The horizontal rows are called **periods** and these are numbered 1–7 going down the Periodic Table.

Between Groups II and III is the block of elements known as the **transition elements** (Figure 9.4).

The Periodic Table can be divided into two as shown by the bold line that starts beneath boron, in Figure 9.3. The elements to the left of this line are metals (fewer than three-quarters) and those on the right are non-metals (fewer than one-quarter). The elements which lie on this dividing line are known as **metalloids** (Figure 9.5). The metalloids behave in some ways as metals and in others as non-metals.

If you look at the properties of the elements across a period of the Periodic Table, you will notice certain trends. For example, there is:

» a gradual change from metal to non-metal
» an increase in the number of electrons in the outer electron shell of the element
» a change in the structure of the element, from giant metallic in the case of metals (e.g. magnesium, p. 50, Figure 3.34), through giant covalent (e.g. diamond, p. 46, Figure 3.29), to simple molecular (e.g. chlorine, p. 39, Figure 3.13).

▲ **Figure 9.5** The metalloid silicon is used to make silicon 'chips'

9.2 Electronic configuration and the Periodic Table

The number of electrons in the outer electron shell is discussed in Chapter 2 (p. 26). This corresponds with the number of the group in the Periodic Table in which the element is found. For example, the elements shown in Table 9.1 have one electron in their outer electron shell and they are all found in Group I. The elements in Group 0, however, are an exception to this rule, as they have two or eight electrons in their outer electron shell. The outer electrons are mainly responsible for the chemical properties of any element and, therefore, elements in the same group have similar chemical properties (Tables 9.2 and 9.3). The number of occupied shells is equal to the period number in which that element is found. For example, we know that sodium is in

Group I because it has only one electron in its outer shell but we can also say that it is in Period 3 of the Periodic Table as it has electrons in the first three shells.

▼ **Table 9.1** Electronic configuration of the first three elements of Group I

Element	Symbol	Proton number	Electronic configuration
Lithium	Li	3	2,1
Sodium	Na	11	2,8,1
Potassium	K	19	2,8,8,1

▼ **Table 9.2** Electronic configuration of the first three elements of Group II

Element	Symbol	Proton number	Electronic configuration
Beryllium	Be	4	2,2
Magnesium	Mg	12	2,8,2
Calcium	Ca	20	2,8,8,2

▼ **Table 9.3** Electronic configuration of the first three elements in Group VII

Element	Symbol	Proton number	Electronic configuration
Fluorine	F	9	2,7
Chlorine	Cl	17	2,8,7
Bromine	Br	35	2,8,18,7

The metallic character of the elements in a group increases as you move down the group. This is because electrons become easier to lose as the outer shell electrons become further from the nucleus. There is less attraction between the nucleus and the outer shell electrons because of the increased distance between them.

9.3 Group I – the alkali metals

Group I consists of the five metals lithium, sodium, potassium, rubidium and caesium, and the radioactive element francium. Lithium, sodium and potassium are commonly available for use in school. They are all very reactive metals and they are stored under oil to prevent them coming into contact with water or air. These three metals have the following properties.

» They are good conductors of electricity and heat.
» They are soft metals. Lithium is the hardest and potassium the softest.
» They are metals with low densities. For example, lithium has a density of 0.53 g/cm³ and potassium has a density of 0.86 g/cm³.
» They have shiny surfaces when freshly cut with a knife (Figure 9.6).

▲ **Figure 9.6** Cutting sodium metal

a Potassium reacts very vigorously with cold water

b An alkaline solution is produced when potassium reacts with water

▲ **Figure 9.7**

» They have low melting points. For example, lithium has a melting point of 181°C and potassium has a melting point of 64°C.
» They burn in oxygen or air, with characteristic flame colours, to form white solid oxides. For example, lithium reacts with oxygen in air to form white lithium oxide, according to the following equation:

lithium + oxygen → lithium oxide

$$4Li(s) + O_2(g) \rightarrow 2Li_2O(s)$$

These Group I oxides all react with water to form alkaline solutions of the metal hydroxide.

lithium oxide + water → lithium hydroxide

$$Li_2O(s) + H_2O(l) → 2LiOH(aq)$$

» They react vigorously with water to give an alkaline solution of the metal hydroxide as well as producing hydrogen gas. For example:

potassium + water → potassium + hydrogen gas
                              hydroxide

$$2K(s) + 2H_2O(l) → 2KOH(aq) + H_2(g)$$

» Of the first three metals in Group I, potassium is the most reactive towards water (Figure 9.7), followed by sodium and then lithium. Such gradual changes we call **trends**. Trends are useful to chemists as they allow predictions to be made about elements we have not observed in action.

» They react vigorously with halogens, such as chlorine, to form metal halides, for example, sodium chloride (Figure 9.8).

sodium + chlorine → sodium chloride

$$2Na(s) + Cl_2(g) → 2NaCl(s)$$

Considering the group as a whole, the further down the group you go, the more reactive the metals become. Francium is, therefore, the most reactive Group I metal.

Table 9.1 shows the electronic configuration of the first three elements of Group I. You will notice in each case that the outer electron shell contains only one electron. When these elements react, they lose this outer electron and, in doing so, become more stable because they obtain the electronic configuration of a noble gas. You will learn more about the stable nature of these gases later in this chapter.

When, for example, the element sodium reacts, it loses its outer electron. This requires energy to overcome the electrostatic attractive forces between the outer electron and the positive nucleus (Figure 9.9).

▲ **Figure 9.8** A very vigorous reaction takes place when sodium burns in chlorine gas. Sodium chloride is produced

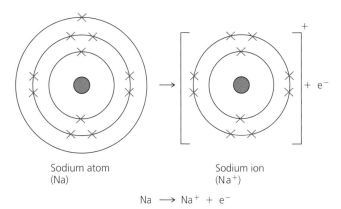

Sodium atom (Na)   Sodium ion (Na$^+$)

$$Na → Na^+ + e^-$$

▲ **Figure 9.9** This sodium atom loses an electron to become a sodium ion

Look at Figure 9.10. Why do you think potassium is more reactive than lithium or sodium?

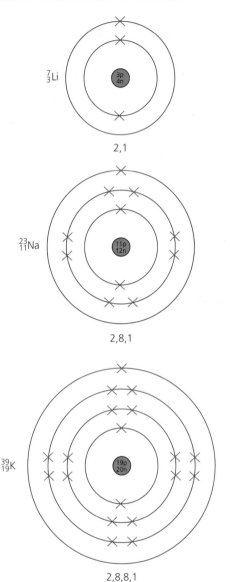

$^{7}_{3}Li$

2,1

$^{23}_{11}Na$

2,8,1

$^{39}_{19}K$

2,8,8,1

▲ **Figure 9.10** Electronic configuration of lithium, sodium and potassium

Potassium is more reactive because less energy is required to remove the outer electron from its atom than for lithium or sodium. This is because as you go down the group, the size of the atoms increases and the outer electron gets further away from the nucleus and becomes easier to remove.

Test yourself

1 Write word and balanced chemical equations for the reactions between:
   a sodium and oxygen
   b sodium and water.
2 a Using the information on p. 139, predict the properties of the element francium related to its melting point, density and softness.
   b Predict how francium would react with water and write a balanced equation for the reaction.
3 Write word and balanced chemical equations for the reactions between:
   a magnesium and water
   b calcium and oxygen.
4 Explain the fact that calcium is more reactive than magnesium in terms of their electronic configurations.

➡ Going further

Group II – the alkaline earth metals

Group II consists of the five metals beryllium, magnesium, calcium, strontium and barium, and the radioactive element radium. Magnesium and calcium are generally available for use in school. These metals have the following properties.

● They are harder than those in Group I.
● They are silvery-grey in colour when pure and clean. They tarnish quickly, however, when left in air due to the formation of a metal oxide on their surfaces (Figure 9.11).

▲ **Figure 9.11** Tarnished (left) and cleaned-up magnesium

● They are good conductors of heat and electricity.
● They burn in oxygen or air with characteristic flame colours to form solid white oxides. For example:

magnesium + oxygen → magnesium oxide

$$2Mg(s) \quad + \quad O_2(g) \quad \rightarrow \quad 2MgO(s)$$

- They react with water, but they do so much less vigorously than the elements in Group I. For example:

$$\text{calcium} + \text{water} \rightarrow \text{calcium hydroxide} + \text{hydrogen gas}$$

$$Ca(s) + 2H_2O(l) \rightarrow Ca(OH)_2(aq) + H_2(g)$$

Considering the group as a whole, the further down the group you go, the more reactive the elements become.

9.4 Group VII – the halogens

Group VII consists of the four elements fluorine, chlorine, bromine and iodine, and the radioactive element astatine. Of these five elements, chlorine, bromine and iodine are generally available for use in school.

>> These elements are coloured and become darker going down the group (Table 9.4).
>> They exist as diatomic molecules, for example, Cl_2, Br_2 and I_2.
>> At room temperature and pressure they show a gradual change from a gas (Cl_2), through a liquid (Br_2), to a solid (I_2) (Figure 9.12) as the density increases.
>> They form molecular compounds with other non-metallic elements, for example HCl.
>> They react with hydrogen to produce the hydrogen halides, which dissolve in water to form acidic solutions.

hydrogen + chlorine → hydrogen chloride

$$H_2(g) + Cl_2(g) \rightarrow 2HCl(g)$$

hydrogen chloride + water → hydrochloric acid

$$HCl(g) + H_2O \rightarrow HCl(aq) \rightarrow H^+(aq) + Cl^-(aq)$$

>> They react with metals to produce ionic metal halides, for example, chlorine and iron produce iron(III) chloride.

iron + chlorine → iron(III) chloride

$$2Fe(s) + 3Cl_2(g) \rightarrow 2FeCl_3(s)$$

a   Chlorine, bromine and iodine

b   Chlorine gas bleaches moist indicator paper
▲ **Figure 9.12**

▼ **Table 9.4** Colours of some halogens

Halogen	Colour
Chlorine	Pale yellow-green gas
Bromine	Red-brown liquid
Iodine	Grey-black solid

Displacement reactions

If chlorine is bubbled into a solution of potassium iodide, the less reactive halogen, iodine, is displaced by the more reactive halogen, chlorine, as you can see from Figure 9.13:

potassium + chlorine → potassium + iodine
iodide   chloride

$$2KI(aq) + Cl_2(g) \rightarrow 2KCl(aq) + I_2(aq)$$

▲ **Figure 9.13** Iodine being displaced from potassium iodide solution as chlorine is bubbled through

The observed order of reactivity of the halogens, confirmed by similar **displacement reactions**, is:

$$\xrightarrow{\text{decreasing reactivity}}$$
chlorine   bromine   iodine

You will notice that, in contrast to the elements of Groups I and II, the order of reactivity decreases on going down the group.

Table 9.5 shows the electronic configuration for chlorine and bromine. In each case, the outer electron shell contains seven electrons. When these elements react, they gain one electron per atom to gain the stable electronic configuration of a noble gas. You will learn more about the stable nature of these gases in the next section. For example, when chlorine reacts, it gains a single electron and forms a negative ion (Figure 9.14).

▼ **Table 9.5** Electronic configuration of chlorine and bromine

Element	Symbol	Proton number	Electronic configuration
Chlorine	Cl	17	2,8,7
Bromine	Br	35	2,8,18,7

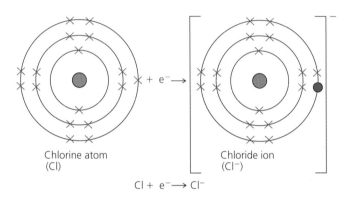

$$Cl + e^- \rightarrow Cl^-$$

▲ **Figure 9.14** A chlorine atom gains an electron to form a chloride ion

Chlorine is more reactive than bromine because the incoming electron is gained more easily by the smaller chlorine atom than in the larger bromine atom. It is gained more easily because there is a stronger attraction between the negative charge of the incoming electron and the positive charge of the nucleus. In the larger bromine atom, there are more occupied electron shells surrounding the nucleus, which lessen the attraction of the nucleus, and the electrons in these shells repel the incoming electron. This makes it harder for the bromine atom to gain the extra electron it needs to gain a stable electronic configuration. This is the reason the reactivity of the halogens decreases going down the group.

The halogens and their compounds are used in many different ways (Figure 9.15).

In the reaction of chlorine with potassium iodide, both Cl atoms in Cl_2 gain an electron from an iodide ion, I^-, thus forming two chloride ions, Cl^-. The iodine atoms formed by the loss of an electron combine to give an iodine molecule, I_2.

Oxidation: $2I^-(aq) + Cl_2(g) \rightarrow 2Cl^-(aq) + I_2(aq)$

Number: –1   0   –1   0

Colour: colourless   green   colourless   orange-brown

The iodide ion has been oxidised because it has lost electrons. The oxidation number has increased. Chlorine has been reduced because it has gained electrons. The oxidation number has decreased.

Going further

Uses of the halogens

- Fluorine is used in the form of fluorides in drinking water and toothpaste because it reduces tooth decay by hardening the enamel on teeth.
- Chlorine is used to make PVC plastic as well as household bleaches. It is also used to kill bacteria and viruses in drinking water (Chapter 11, p. 174).
- Bromine is used to make disinfectants, medicines and fire retardants.
- Iodine is used in medicines and disinfectants, and also as a photographic chemical.

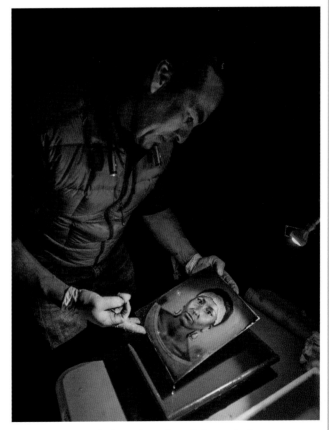

▲ **Figure 9.15** The halogens have many varied uses – fluoride in toothpaste to help reduce dental decay, iodine as a photographic chemical, chlorine in household bleach to kill bacteria, and bromine as a fire retardant

Practical skills

Halogen displacement reactions

For safe experiments/demonstrations which are related to this chapter, please refer to the *Cambridge IGCSE Chemistry Practical Skills Workbook*, which is also part of this series.

Safety

- Eye protection must be worn.

Displacement reactions can be used to determine the reactivity of the Group VII elements, the halogens. In this experiment, a student uses these reactions to determine an order of reactivity for iodine, chlorine and bromine, and predicts the reactivity of the other two halogens, fluorine and astatine.

If a more reactive halogen reacts with a compound of a less reactive halogen, the less reactive halogen will be displaced and it will form the halogen molecule, while the more reactive halogen becomes a halide ion.

To observe the presence of the different halogens in solution, the student used an organic solvent as the halogens produce more vivid colours in this solvent compared with water.

To show the colour of the halogens in the organic solvent solutions, the student separately placed three halogens in test tubes and added a small amount of the organic solvent. The tubes were then fitted with a rubber bung and shaken.

After shaking, two layers formed. The lower layer is the water (aqueous layer) and the upper layer is the organic layer.

▲ **Figure 9.16** The tube on the left shows the results of chlorine water being added to sodium bromide, and on the right the result of bromine water being added to sodium iodide

The results obtained are:

Halogen	Colour in water	Colour in organic solvent
Chlorine	Colourless	Pale green
Bromine	Orange	Orange
Iodine	Brown	Violet

The colours of the halogens in water can be used to identify if a reaction has occurred when a solution of a halogen, for example, chlorine water, is mixed with a solution of a sodium halide, for example, sodium iodide solution. The table below shows the results of six of these.

Sample data

		Chlorine water	Bromine water	Iodine water
Colour after shaking with sodium iodide solution		Brown	Brown	
Colour of each layer with the organic solvent added	Upper	Violet	Violet	
	Lower	Brown	Brown	
Colour after shaking with sodium bromide solution		Orange		Brown
Colour of each layer with the organic solvent added	Upper	Orange		Violet
	Lower	Pale orange		Brown
Colour after shaking with sodium chloride solution			Orange	Brown
Colour of each layer with the organic solvent added	Upper		Orange	Violet
	Lower		Pale orange	Brown

1 In how many of the six reactions has a displacement reaction occurred?
2 For each of these displacement reactions, write a balanced chemical equation.
3 What type of chemical change has happened to the halogen molecules in the displacement reactions?
4 State the order of reactivity of the three halogens used in the experiment.
5 What would be the colour of fluorine and astatine in water?
6 Which of the five halogen elements is the most reactive? Explain your answer.
7 Why were solutions of the halogens used, rather than the halogens themselves?

9.5 Group VIII – the noble gases

Helium, neon, argon, krypton, xenon and the radioactive element radon make up a most unusual group of non-metals, called the noble gases. They were all discovered after Mendeleev had published his Periodic Table. They were discovered between 1894 and 1900, mainly through the work of the British scientists Sir William Ramsay and Lord John William Strutt Rayleigh.

» They are colourless gases.
» They are monatomic gases – they exist as individual atoms, for example, He, Ne and Ar.
» They are very unreactive.

No compounds of helium, neon or argon have ever been found. However, more recently a number of compounds of xenon and krypton with fluorine and oxygen have been produced, for example, XeF_6.

These gases are chemically unreactive because they have electronic configurations which are stable and very difficult to change (Table 9.6). They are so stable that other elements attempt to attain these electronic configurations during chemical reactions (Chapter 3, p. 31, and p. 38). You have probably seen this in your study of the elements of Groups I, II and VII.

▼ **Table 9.6** Electronic configuration of helium, neon and argon

Element	Symbol	Proton number	Electronic configuration
Helium	He	2	2
Neon	Ne	10	2,8
Argon	Ar	18	2,8,8

Although unreactive, they have many uses. Argon, for example, is the gas used to fill light bulbs to prevent the tungsten filament reacting with air. Neon is used extensively in advertising signs and in lasers.

> ## Test yourself
> 5 Write word and balanced chemical equations for the reactions between:
> a bromine and potassium iodide solution
> b bromine and potassium chloride solution.
> If no reaction will take place, write 'no reaction' and explain why.
> c Using the information on pp. 141–142, predict the properties of the element astatine related to its melting point, density and physical state at room temperature. Predict how astatine would react with sodium bromide solution.

9.6 Transition elements

This block of metals includes many you will be familiar with, for example, copper, iron, nickel, zinc and chromium (Figure 9.17).

» They are less reactive metals.
» They form a range of brightly coloured compounds (Figure 9.18).
» They are harder and stronger than the metals in Groups I and II.
» They have much higher densities than the metals in Groups I and II.
» They have high melting points (except for mercury, which is a liquid at room temperature).
» They are good conductors of heat and electricity.
» They show catalytic activity (Chapter 7, p. 106) as elements and compounds. For example, iron is used in the industrial production of ammonia gas (Haber process, Chapter 7, p. 109).
» They do not react (corrode) so quickly with oxygen and/or water.

a Copper is used in many situations which involve good heat and electrical conduction. It is also used in medallions and bracelets

b These gates are made of iron. Iron can easily be moulded into different shapes

d This bucket has been coated with zinc to prevent the steel of the bucket corroding

c Monel is an alloy of nickel and copper. It is extremely resistant to corrosion, even that caused by sea water

e The alloy stainless steel contains a high proportion of chromium, which makes it corrosion resistant

▲ **Figure 9.17** Everyday uses of transition elements and their compounds. They are often known as the 'everyday metals' a   Some solutions of coloured transition element compounds

▲ **Figure 9.18**

b   The coloured compounds of transition elements can be seen in these pottery glazes

» They form simple ions with variable oxidation numbers. (For a discussion of oxidation number, see Chapter 3, p. 35.) For example, copper forms Cu^+ (Cu(I)) and Cu^{2+} (Cu(II)), in compounds such as Cu_2O and $CuSO_4$, and iron forms Fe^{2+} (Fe(II)) and Fe^{3+} (Fe(III)), in compounds such as $FeSO_4$ and $FeCl_3$.

» They form more complicated ions with high oxidation numbers. For example, chromium forms the dichromate(VI) ion, $Cr_2O_7^{2-}$, which contains chromium with a +6 oxidation state (Cr(VI)) and manganese forms the manganate(VII) ion, MnO_4^-, which contains manganese with a +7 oxidation number (Mn(VII)).

▼ **Table 9.7** Comparison of hydrogen with lithium and fluorine

Lithium	Hydrogen	Fluorine
Solid	Gas	Gas
Forms a positive ion	Forms positive or negative ions	Forms a negative ion
1 electron in outer electron shell	1 electron in outer electron shell	1 electron short of a full outer electron shell
Loses 1 electron to form a noble gas configuration	Needs 1 electron to form a noble gas configuration	Needs 1 electron to form a noble gas configuration

Test yourself

6   Look at the photographs in Figure 9.17 and state which key properties are important when considering the particular use of the metal.
7   Which groups in the Periodic Table contain:
  a   only metals?
  b   only non-metals?
  c   both metals and non-metals?

9.7 The position of hydrogen

Hydrogen is often placed by itself in the Periodic Table. This is because the properties of hydrogen are unique. However, useful comparisons can be made with the other elements. It is often shown at the top of either Group I or Group VII, but it cannot fit easily into the trends shown by either group; see Table 9.7.

Revision checklist

After studying Chapter 9 you should be able to:
✓   Use the Periodic Table to help you to predict the properties of elements.
✓   Use the Periodic Table to help you to give the charges on ions and to write chemical and ionic equations.
✓   Describe the trends in the properties and reactions of the Group I and Group VII elements.
✓   Describe the Group VIII elements as unreactive, monoatomic gases and explain why they are unreactive.

✓   Identify trends in the different groups if you are given information.

✓   Describe the physical appearance of the Group I and VII elements.
✓   Describe halogen-halide displacement reactions.
✓   Give the properties of the transition metals.

✓   Understand that transition metal ions have variable oxidation numbers.

Exam-style questions

1 The diagram below shows part of the Periodic Table.

I	II								III	IV	V	VI	VII	0
		H												He
Li	Be								B			O		
										Si	P		Cl	Ar
	Ca								Zn	Ga			Se	Br

Using **only** the symbols of the elements shown above, give the symbol for an element which:
a is a pale yellow-green coloured toxic gas [1]
b is stored under oil [1]
c has five electrons in its outer electron energy shell [1]
d is the most reactive Group II element [1]
e is the most reactive halogen [1]
f is the only liquid shown [1]
g is a transition element [1]
h is a gas with two electrons in its outer shell. [1]

2 Three members of the halogens are $^{35.5}_{17}Cl$, $^{80}_{35}Br$ and $^{127}_{53}I$.
a i Give the electronic configuration of an atom of chlorine. [1]
ii Explain why the relative atomic mass of chlorine is not a whole number. [1]
iii Identify how many protons there are in an atom of bromine. [1]
iv Identify how many neutrons there are in an atom of iodine. [1]
v Explain the order of reactivity of these elements. [5]
b When potassium is allowed to burn in a gas jar of chlorine, in a fume cupboard, clouds of white smoke are produced.
i Explain why this reaction is carried out in a fume cupboard. [1]
ii State what the white smoke consists of. [1]
iii Give a word and balanced chemical equation for this reaction. [2]
iv Describe what you would expect to see when potassium is allowed to burn safely in a gas jar of bromine vapour. Write a word and balanced chemical equation for this reaction. [3]

3 'By using displacement reactions, it is possible to deduce the order of reactivity of the halogens.' Discuss this statement with reference to the elements bromine, iodine and chlorine only. [4]
4 Use the information given in the table below to answer the questions concerning the elements **Q, R, S, T** and **X**.

Element	Proton number	Mass number	Electronic configuration
Q	3	7	2,1
R	20	40	2,8,8,2
S	18	40	2,8,8
T	8	18	2,6
X	19	39	2,8,8,1

a Identify the element that has 22 neutrons in each atom. [1]
b Identify the element that is a noble gas. [1]
c Identify the two elements that form ions with the same electronic configuration as argon. [2]
d Identify the two elements that are in the same group of the Periodic Table and which group this is. [2]
e Place the elements in the table into the periods in which they belong. [3]

f Identify the most reactive metal element in the table. [1]

g Identify which of the elements is calcium. [1]
5 a Consider the chemical properties and physical properties of the halogens chlorine, bromine and iodine. Using these properties, predict the following about the other two halogens, fluorine and astatine. [2]

Property	Fluorine	Astatine
State at room temperature and pressure		
Colour		
Reactivity with sodium metal		

b **i** Give a word equation for the reaction of chlorine gas with sodium bromide solution. [1]

ii Give a balanced chemical equation for the reaction, with state symbols. [2]

iii Give an ionic equation for the reaction, with state symbols. [2]

6 Some of the most important metals we use are found in the transition element section of the Periodic Table. One of these elements is copper. Sodium, a Group I metal, has very different properties from those of copper. Complete the table below to show their differences. [4]

	Transition element, e.g. copper	Group I metal, e.g. sodium
Hardness (hard/soft)		
Reactivity		
Density (high/low)		
Variable oxidation states		

Metals

In this chapter, you will look at the extraction, properties and chemical reactions of metals. Some of these metallic elements are the most important elements in the world. They are used in the building of roads, bridges, factories, in the movement of electricity and as catalysts. With all these metals available to use, what properties make some better than others for a particular job?

We will look at the way in which we are able to modify the properties of metals to make them have different and better properties and how we try to protect metallic objects from corrosion to extend their useful life. Why is this important?

By the end of this chapter, you should be able to predict the chemical properties of a metal to decide which metal would be best for a particular job.

10.1 Properties of metals

The majority of the elements in the Periodic Table are metals. They have very different properties to the non-metallic elements. A comparison of their properties is shown in Table 10.1.

▼ **Table 10.1** How the properties of metals and non-metals compare

Property	Metal	Non-metal
Thermal conductivity	Good	Poor
Electrical conductivity	Good	Poor
Malleability	Good	Poor – usually soft or brittle
Ductility	Good	Poor – usually soft or brittle
Melting point	Usually high	Usually low
Boiling point	Usually high	Usually low

You have already seen in Chapter 2, p. 11, that metals usually have similar physical properties. However, they differ in other ways. Look closely at the three photographs in Figure 10.1.

a Sodium burning in air/oxygen

b Iron rusts when left unprotected

c   Gold is used in leaf form on this giant Buddha as it is unreactive

▲ **Figure 10.1**

Sodium is soft and reacts violently with both air and water. Iron also reacts with air and water but much more slowly, forming **rust**. Gold, however, remains totally unchanged after many hundreds of years. Sodium is said to be more reactive than iron and, in turn, iron is said to be more reactive than gold.

10.2 Metal reactions

By carrying out reactions in the laboratory with other metals and with air, water and dilute acid, it is possible to produce an order of reactivity of the metals.

With acid

If a metal reacts with dilute hydrochloric acid then hydrogen and the metal chloride are produced. Look closely at the photograph in Figure 10.2 showing magnesium metal reacting with dilute hydrochloric acid. You will notice effervescence, which is caused by bubbles of hydrogen gas being formed as the reaction between the two substances proceeds. The other product of this reaction is the salt, magnesium chloride.

magnesium + hydrochloric → magnesium + hydrogen
            acid   chloride

$$Mg(s) \quad + \quad 2HCl(aq) \quad \rightarrow \quad MgCl_2(aq) + \quad H_2(g)$$

If similar reactions are carried out using other metals with acid, an order of reactivity can be produced by measuring the rate of evolution of hydrogen. This is known as a **reactivity series**.

▲ **Figure 10.2** Effervescence occurs when magnesium is put into acid

> **Key definition**
>
> The **reactivity series** is the order of the reactivity of the following elements: potassium, sodium, calcium, magnesium, aluminium, carbon, zinc, iron, hydrogen, copper, silver, gold.

With air/oxygen

Many metals react directly with oxygen to form oxides. For example, magnesium burns brightly in oxygen to form the white powder magnesium oxide.

magnesium + oxygen → magnesium oxide

$$2Mg(s) \; + \; O_2(g) \; \rightarrow \quad 2MgO(s)$$

With cold water/steam

Reactive metals such as potassium, sodium and calcium react with cold water to produce the metal hydroxide and hydrogen gas. For example, the reaction of sodium with water produces sodium hydroxide and hydrogen.

sodium + water → sodium hydroxide + hydrogen

$$2Na(s) + 2H_2O(l) \rightarrow \quad 2NaOH(aq) \quad + \quad H_2(g)$$

The moderately reactive metals, magnesium, zinc and iron, react slowly with water. They will, however, react more rapidly with steam (Figure 10.3). In their reaction with steam, the metal oxide and hydrogen are formed. For example, magnesium produces magnesium oxide and hydrogen gas.

magnesium + steam → magnesium oxide + hydrogen

$$Mg(s) \quad + H_2O(g) \rightarrow \quad MgO(s) \quad + \quad H_2(g)$$

This experiment should only ever be carried out as a teacher demonstration. Eye protection must be worn during the demonstration. There is the danger of suckback of water into the boiling tube, so your teacher will remove the delivery tube from the water before heating is stopped.

An order of reactivity, giving the most reactive metal first, using results from experiments with dilute acid, is shown in Table 10.2. The table also shows how the metals react with air/oxygen and water/steam and, in addition, the ease of extraction of the metal.

In all these reactions, the most reactive metal is the one that has the highest tendency to lose outer electrons to form a positive metal ion.

▲ **Figure 10.3** Apparatus used to investigate how metals such as magnesium react with steam

▼ **Table 10.2** Order of reactivity

Reactivity series	Reaction with dilute acid	Reaction with air/oxygen	Reaction with water	Ease of extraction	
Potassium (K) Sodium (Na)	Produce H_2 with decreasing vigour	Burn very brightly and vigorously	Produce H_2 with decreasing vigour with cold water	Difficult to extract	Increasing reactivity of metal
Calcium (Ca) Magnesium (Mg)		Burn to form an oxide with decreasing vigour	React with steam with decreasing vigour	Easier to extract	
Aluminium (Al*)					
[Carbon (C)]					
Zinc (Zn)					
Iron (Fe)					
[Hydrogen (H)]		React slowly to form the oxide	Do not react with cold water or steam		
Copper (Cu)	Do not react with dilute acids				
Silver (Ag)		Do not react		Found as the element (uncombined)	
Gold (Au)					

* Because aluminium reacts so readily with the oxygen in the air, a protective oxide layer is formed on its surface. This often prevents any further reaction and disguises aluminium's true reactivity. This gives us the use of a light and strong metal.

a This wood-burning stove is made of iron

b Copper pots and pans

▲ **Figure 10.4**

▲ **Figure 10.5** Planes are made of an alloy which contains magnesium and aluminium

Generally, it is the unreactive metals that we find the most uses for; for example, the metals iron and copper can be found in many everyday objects (Figure 10.4). However, magnesium is one of the metals used in the construction of the Airbus A380 (Figure 10.5).

Aluminium, although a reactive metal, has a protective oxide coating on its surface which forms

when it is manufactured. This allows it to be used in many areas of industry, such as in the manufacture of cars, aircraft and overhead electricity cables, because of its low density and good electrical conductivity.

Both sodium and potassium are so reactive that they have to be stored under oil to prevent them from coming into contact with water or air. However, because they have low melting points and are good conductors of heat, they are used as coolants for nuclear reactors.

> **Test yourself**
>
> 1 Write balanced chemical equations for the reactions between:
>   a iron and dilute hydrochloric acid
>   b potassium and water
>   c zinc and oxygen
>   d calcium and water.
> 2 Make a list of six things you have in your house made from copper or iron.
>   Give a use for each of the other unreactive metals shown in the reactivity series.

10.3 Reactivity of metals and their uses

Generally, it is the unreactive metals for which we find the most uses. For example, copper is used for electrical wiring because of its good electrical conductivity and ductility. However, the metal aluminium is an exception. Aluminium appears in the reactivity series just below magnesium and is quite reactive.

Fortunately, it forms a relatively thick oxide layer on the surface of the metal which prevents further reaction. This gives us the use of a light, strong metal for use in:

» the manufacture of aircraft because of its low density
» the manufacture of overhead electrical cables because of its low density and good electrical conductivity
» food containers because of its resistance to corrosion.

> **Test yourself**
>
> 3 Compare the uses of magnesium, a reactive metal, and lead, an unreactive metal in terms of their reactivity and properties.

Displacement reactions

Metals compete with each other for other anions, in solution. This type of reaction is known as a displacement reaction. As in the previous type of competitive reaction, the reactivity series can be used to predict which of the metals will 'win'.

In a displacement reaction, a more reactive metal will displace a less reactive metal from a solution of its salt. Zinc is above copper in the reactivity series. Figure 10.6 shows what happens when a piece of zinc metal is left to stand in a solution of copper(II) nitrate. The copper(II) nitrate slowly loses its blue colour as the zinc continues to displace the copper from the solution and eventually becomes colourless zinc nitrate.

$$\text{zinc} + \text{copper(II) nitrate} \rightarrow \text{zinc nitrate} + \text{copper}$$

$$Zn(s) + Cu(NO_3)_2(aq) \rightarrow Zn(NO_3)_2(aq) + Cu(s)$$

The ionic equation for this reaction is:

$$\text{zinc} + \text{copper ions} \rightarrow \text{zinc ions} + \text{copper}$$

$$Zn(s) + Cu^{2+}(aq) \rightarrow Zn^{2+}(aq) + Cu(s)$$

This is also a redox reaction involving the transfer of two electrons from the zinc metal to the copper ions. The zinc is oxidised to zinc ions in aqueous solution, while the copper ions are reduced. (See Chapter 5, p. 71, for a discussion of oxidation and reduction in terms of electron transfer.) It is possible to confirm the reactivity series for metals using **competition reactions** of the types discussed in this section.

▲ **Figure 10.6** Zinc displaces copper

Practical skills

Using displacement reactions to determine the order of reactivity

For safe experiments/demonstrations which are related to this chapter, please refer to the *Cambridge IGCSE Chemistry Practical Skills Workbook*, which is also part of this series.

Safety

- Eye protection must be worn.

In a displacement reaction a more reactive metal will displace a less reactive metal from a solution of its salt. It is possible to carry out displacement reactions using a dimple tile (Figure 10.7).

▲ **Figure 10.7** Dimple tile

Small pieces of metals can be placed into the dimples and then, using a dropping pipette, a solution containing another metal ion can be put over the metal. It is usual to use metal nitrate solutions as they are all soluble.

After the metal solution has been added to the metal, you must ensure that the metal is below the surface of the solution and leave time for the reaction to occur: usually 2 or 3 minutes is adequate.

1 Why is it important that the metal is under the surface of the solution?
2 What would be observed if a reaction occurred?

Sample data

Five metals and nitrate solutions of those metals were reacted together using the method above. The five metals were tin, magnesium, zinc, copper and iron. The results obtained were collected in the table below. A tick (✓) indicates a colour change was observed, a cross (x) indicates no colour change was observed.

	Nitrates				
	Copper	**Zinc**	**Tin**	**Magnesium**	**Iron**
Copper	-	x	x	x	x
Zinc	✓	-	✓	x	✓
Tin	✓	x	-	x	x
Magnesium	✓	✓	✓	-	✓
Iron	✓	x	✓	x	-

3 Using the information in the table, state which of the metals was the most reactive.
4 Using the information in the table, state which of the metals was the least reactive.
5 Arrange the five metals in order of increasing reactivity.
6 a Write a word, balanced chemical and ionic equation for the reaction between magnesium metal and copper(II) nitrate.
   b What would you observe when this reaction occurs.?

4 Predict whether the following reactions will take place:
a magnesium + calcium nitrate solution
b iron + copper(II) nitrate solution
c copper + silver nitrate solution.
Complete the word equations and write balanced chemical and ionic equations for those reactions which do take place.

5 How could you carry out a series of reactions between metals and solutions of their nitrates to establish a reactivity series?

6 The data below was obtained by carrying out displacement reactions of five metals with the nitrates of the same five metals. Strips of each metal were placed in solutions of the other four metals nitrate solutions.

	Nitrates				
	A	**B**	**C**	**D**	**E**
A	—	✓	✓	✓	
B	x	—	x	x	x
C	x	✓	—	✓	x
D	x	✓	x	—	x
E	✓	✓	✓	✓	—

✓ = metal displaced   x = no reaction
Put the five metals **A–E** in order of their reactivity using the data above.

10.4 Identifying metal ions

When an alkali dissolves in water, it produces hydroxide ions. It is known that most metal hydroxides are insoluble. So if hydroxide ions from a solution of an alkali are added to a solution of a metal salt, an insoluble, often coloured, metal hydroxide is precipitated from solution (Figure 10.8).

If we take the example of iron(III) chloride with sodium hydroxide solution:

iron(III) chloride	+	sodium hydroxide	→	iron(III) hydroxide	+	sodium chloride

$$FeCl_3(aq) + 3NaOH(aq) \rightarrow Fe(OH)_3(s) + 3NaCl(aq)$$

a Iron(III) hydroxide is precipitated

b Copper(II) hydroxide is precipitated

▲ **Figure 10.8**

The ionic equation for this reaction is:

iron(III) ions + hydroxide ions → iron(III) hydroxide

$$Fe^{3+}(aq) + 3OH^-(aq) \rightarrow Fe(OH)_3(s)$$

Table 10.3 shows the effects of adding a few drops of sodium hydroxide solution to solutions containing various metal ions, and of adding an excess. The colours of the insoluble metal hydroxides can be used to identify the metal cations present in solution. In some cases, the precipitate dissolves in excess hydroxide, owing to the amphoteric nature of the metal hydroxide. This amphoteric nature can also be used to help identify metals such as aluminium and zinc.

▼ **Table 10.3** The effect of adding sodium hydroxide solution to solutions containing various metal ions

Metal ion present in solution	Effect of adding sodium hydroxide solution	
	A few drops	**An excess**
Aluminium, Al^{3+}	White precipitate of aluminium hydroxide	Precipitate is soluble in excess, giving a colourless solution
Calcium, Ca^{2+}	White precipitate of calcium hydroxide	Precipitate is insoluble in excess
Chromium(III), Cr^{3+}	Green precipitate of chromium(III) hydroxide	Precipitate is insoluble in excess
Copper(II), Cu^{2+}	Light blue precipitate of copper(II) hydroxide	Precipitate is insoluble in excess
Iron(II), Fe^{2+}	Green precipitate of iron(II) hydroxide	Precipitate is insoluble in excess, turns brown near the surface on standing
Iron(III), Fe^{3+}	Red-brown precipitate of iron(III) hydroxide	Precipitate is insoluble in excess
Zinc, Zn^{2+}	White precipitate of zinc hydroxide	Precipitate is soluble in excess, giving a colourless solution

Test yourself

7 Write ionic equations for the reactions which take place to produce the metal hydroxides shown in Table 10.3.

8 Describe what you would see when sodium hydroxide is added slowly to a solution containing iron(II) nitrate.

10.5 Extraction of metals

Metals have been used since prehistoric times. Many primitive iron tools have been excavated. These were probably made from small amounts of native iron found in rock from meteorites. It was not until about 2500BC that iron became more widely used. This date marks the dawn of the Iron Age, when people learned how to get iron from its ores in larger quantities by reduction using charcoal. An ore is a naturally occurring mineral from which a metal can be extracted. Metals lower in the reactivity series (such as iron and copper) tend to be easier to extract from their ores than more reactive metals (such as potassium and sodium).

The majority of metals are too reactive to exist on their own in the Earth's crust, and they occur naturally in rocks as compounds in ores (Figure 10.9). These ores are usually carbonates, oxides or sulfides of the metal, mixed with impurities.

Some metals, such as gold and silver, occur in a native form as the free metal (Figure 10.10). They are very unreactive and have withstood the action of water and the atmosphere for many thousands of years without reacting to become compounds.

Some of the common ores are shown in Table 10.4.

▲ **Figure 10.10** Gold crystals

Large lumps of the ore are first crushed and ground up by very heavy machinery. Some ores are already fairly concentrated when mined. For example, in some parts of the world, hematite contains over 80% Fe_2O_3. However, other ores, such as copper pyrites, are often found to be less concentrated, with only 1% or less of the copper compound, and so they have to be concentrated before the metal can be extracted. The method used to extract the metal from its ore depends on the position of the metal in the reactivity series.

▲ **Figure 10.9** Chalcopyrite, an ore of copper (top) and galena, an ore of lead (bottom)

▼ **Table 10.4** Some common ores

Metal	Name of ore	Chemical name of compound in ore	Formula	Usual method of extraction
Aluminium	Bauxite	Aluminium oxide	$Al_2O_3.2H_2O$	Electrolysis of oxide dissolved in molten cryolite
Copper	Copper pyrites	Copper iron sulfide	$CuFeS_2$	The sulfide ore is roasted in air
Iron	Hematite	Iron(III) oxide	Fe_2O_3	Heat oxide with carbon
Sodium	Rock salt	Sodium chloride	$NaCl$	Electrolysis of molten sodium chloride
Zinc	Zinc blende	Zinc sulfide	ZnS	Sulfide is roasted in air and the oxide produced is heated with carbon

Because reactive metals, such as sodium, hold on to the element(s) they have combined with, they are usually difficult to extract. For example, sodium chloride (as rock salt) is an ionic compound with the Na^+ and Cl^- ions strongly bonded to one another. The separation of these ions and the subsequent isolation of the sodium metal is therefore difficult.

Electrolysis of the molten, purified ore is the method used in these cases. During this process, the metal is produced at the cathode while a non-metal is produced at the anode. As you might expect, extraction of metal by electrolysis is expensive. In order to keep costs low, many metal smelters using electrolysis are situated in regions where there is hydroelectric power.

For further discussion of the extraction of aluminium, see Chapter 5, p. 72.

Metals towards the middle of the reactivity series, such as iron and zinc, may be extracted by reducing the metal oxide with the non-metal carbon.

Extraction of iron

Iron is extracted mainly from its oxides, hematite (Fe_2O_3) and magnetite (Fe_3O_4), in a blast furnace (Figures 10.11 and 10.12). These ores contain at least 60% iron. The blast furnace is a steel tower, approximately 50 m high, lined with heat-resistant bricks. It is loaded with the 'charge' of iron ore (usually hematite), carbon in the form of coke (made by heating coal) and limestone (calcium carbonate).

A blast of hot air is sent in near the bottom of the furnace through holes which makes the 'charge' glow, as the coke burns in the preheated air.

carbon + oxygen → carbon dioxide

$$C(s) + O_2(g) \rightarrow CO_2(g)$$

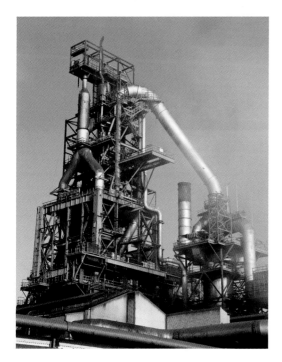

▲ **Figure 10.11** A blast furnace

Labels: Hot waste gases – these are used to heat air going into the blast furnace; Skip dumps charge of iron ore, coke and limestone into hopper; Hopper; Height approximately 90 m; 50 m; Heat-resistant brick; 650°C; 1000°C; 1100°C; 50 rows of water-cooled copper coolers in the lining of the furnace; Hot air blown through tuyère; Slag notch; Molten iron; Molten slag; Tap hole; Firebrick pad

▲ **Figure 10.12** Cross-section of a blast furnace

A number of chemical reactions then follow.

➤➤ The limestone begins to decompose:

calcium carbonate → calcium oxide + carbon dioxide

$$CaCO_3(s) \rightarrow CaO(s) + CO_2(g)$$

➤➤ The carbon dioxide gas produced reacts with more hot coke higher up in the furnace, producing carbon monoxide in an endothermic reaction.

carbon dioxide + coke → carbon monoxide

$$CO_2(g) + C(s) \rightarrow 2CO(g)$$

➤➤ Carbon monoxide is a reducing agent (Chapter 3, p. 37). It rises up the furnace and reduces the iron(III) oxide ore. This takes place at a temperature of around 700°C:

iron(III) + carbon monoxide → iron + carbon dioxide
oxide

$$Fe_2O_3(s) + \quad 3CO(g) \quad \rightarrow 2Fe(s) + \quad 3CO_2(g)$$

The molten iron trickles to the bottom.

• Calcium oxide is a base and this reacts with acidic impurities such as silicon(IV) oxide in the iron, to form a slag which is mainly calcium silicate.

calcium oxide + silicon(IV) oxide → calcium silicate

$$CaO(s) \quad + \quad SiO_2(s) \quad \rightarrow \quad CaSiO_3(s)$$

The slag trickles to the bottom of the furnace, but because it is less dense than the molten iron, it floats on top of it.

Generally, metallic oxides, such as calcium oxide (CaO), are basic whereas non-metallic oxides, such as silicon(IV) oxide (SiO_2), are acidic.

Certain oxides, such as carbon monoxide (CO), are neutral and others, such as zinc oxide (ZnO), are amphoteric.

The molten iron, as well as the molten slag, may be tapped off (run off) at regular intervals.

The waste gases, mainly nitrogen and oxides of carbon, escape from the top of the furnace. They are used in a heat exchange process to heat incoming air and so help to reduce the energy costs of the process. Slag is the other waste material. It is used by builders and road makers (Figure 10.13) for foundations.

▲ **Figure 10.13** Slag is used in road foundations

The extraction of iron is a continuous process and is much cheaper to run than an electrolytic method.

▲ **Figure 10.14** Redcar steelworks at Teesside, UK

▶ Test yourself

9 How does the method used for extracting a metal from its ore depend on the metal's position in the reactivity series?

10 'It is true to say that almost all the reactions by which a metal is extracted from its ore are reduction reactions.' Discuss this statement with respect to the extraction of iron and aluminium.

➡ Going further

Metal waste

Recycling has become commonplace in recent years (Figure 10.15). Why should we really want to recycle metals? Certainly, if we extract fewer metals from the Earth then the existing reserves will last that much longer. Also, recycling metals prevents the creation of a huge environmental problem (Figure 10.16). However, one of the main considerations is that it saves money.

▲ **Figure 10.15** Aluminium can recycling

The main metals which are recycled include aluminium and iron. Aluminium is saved by many households as drinks cans and milk bottle tops, to be melted down and recast. Iron is collected at community tips in the form of discarded household goods and it also forms a large part of the materials collected by scrap metal dealers. Iron is recycled to steel. Many steel-making furnaces run mainly on scrap iron.

Aluminium is especially easy to recycle at low cost. Recycling uses only 5% of the energy needed to extract the metal by electrolysis from bauxite. Approximately 60% of the European need for aluminium is obtained by recycling.

▲ **Figure 10.16** If we did not recycle metals, then this sight would be commonplace

10.6 Metal corrosion

After a period of time, objects made of iron or steel will become coated with rust. The **rusting** of iron is a serious problem and wastes enormous amounts of money each year. Estimates are difficult to make, but it is thought that upwards of £1 billion a year is spent worldwide on replacing iron and steel structures.

Rust is an orange–red powder consisting mainly of hydrated iron(III) oxide ($Fe_2O_3 \cdot xH_2O$). Both water and oxygen are essential for iron to rust, and if one of these two substances is not present then rusting will not take place. The rusting of iron is encouraged by salt. Figure 10.17 (p. 162) shows an experiment to show that oxygen (from the air) and water are needed for iron to rust.

> **Key definition**
> **Rusting** of iron and steel to form hydrated iron(II) oxide requires both water and oxygen.

Rust prevention

To prevent iron rusting, it is necessary to stop oxygen (from the air) and water coming into contact with it. There are several ways of doing this.

Painting

Ships, lorries, cars, bridges and many other iron and steel structures are painted to prevent rusting (Figure 10.18). However, if the paint is scratched, the iron beneath it will start to rust (Figure 10.19) and corrosion can then spread under the paintwork which is still sound. This is why it is essential that the paint is kept in good condition and checked regularly.

Tube 1
(control
experiment)   Tube 2   Tube 3   Tube 4

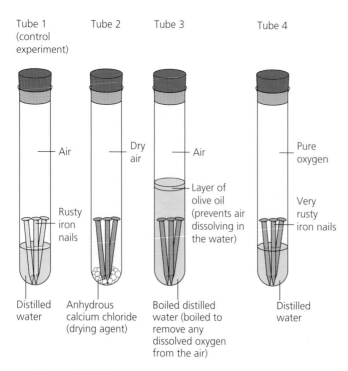

Air   Air

Dry
air

Rusty
iron
nails

Layer of
olive oil
(prevents air
dissolving in
the water)

Pure
oxygen

Very
rusty
iron nails

Distilled
water

Anhydrous
calcium chloride
(drying agent)

Boiled distilled
water (boiled to
remove any
dissolved oxygen
from the air)

Distilled
water

▲ **Figure 10.17** Rusting experiment with nails

▲ **Figure 10.18** Painting keeps the air and water away from the steel used to build a ship

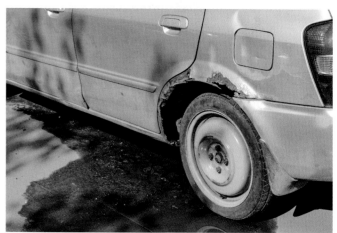

▲ **Figure 10.19** A brand new car is protected against corrosion (top). However, if the paintwork is damaged, then rusting will result

Oiling/greasing

The iron and steel in the moving parts of machinery are coated with oil to prevent them from coming into contact with air or moisture. This is the most common way of protecting moving parts of machinery, but the protective film must be renewed.

Coating with plastic

The exteriors of refrigerators, freezers and many other items are coated with plastic, such as PVC, to prevent the steel structure rusting (Figure 10.20).

▲ **Figure 10.20** A coating of plastic stops metal objects coming into contact with oxygen or water

Galvanising

Some steel beams, used in the construction of bridges and buildings, are galvanised; steel waste collection bins are also galvanised. This involves dipping the object into molten zinc. The thin layer of the more reactive zinc metal coating the steel object slowly corrodes and loses electrons to the iron, thereby protecting it. This process continues even when much of the layer of zinc has been scratched away, so the iron continues to be protected (Figure 10.21).

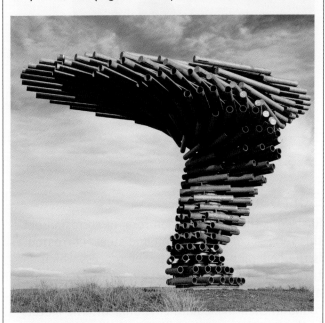

▲ **Figure 10.21** The Burnley Singing Ringing Tree is a sculpture made from galvanised tubes

Sacrificial protection

Bars of zinc are attached to the hulls of ships and to oil rigs (as shown in Figure 10.22a). Zinc is above iron in the reactivity series and will react in preference to it and so is corroded. It forms positive ions more easily than the iron.

$$Zn(s) + Fe^{2+}(aq) \rightarrow Zn^{2+}(aq) + Fe(s)$$

As long as some of the zinc bars remain in contact with the iron structure, the structure will be protected from rusting. When the zinc runs out, it must be renewed. Gas and water pipes made of iron and steel are connected by a wire to blocks of magnesium to obtain the same result. In both cases, as the more reactive metal corrodes, it loses electrons to the iron and so protects it (Figure 10.22b).

a   Bars of zinc on the hull of a ship

b   The zinc is sacrificed to protect the steel. Electrons released from the dissolving zinc cause reduction to occur at the surface of the hull

▲ **Figure 10.22** Sacrificial protection

Corrosion

Rusting is the most common form of corrosion but this term is confined to iron and steel. **Corrosion** is the general name given to the process which takes place when metals and alloys are chemically attacked by oxygen, water or any other substances found in their immediate environment. The metals in the reactivity series will corrode to a greater or lesser extent. Generally, the higher the metal is in the reactivity series, the more rapidly it will corrode. If sodium and potassium were not stored under oil, they would corrode very rapidly indeed. Magnesium, calcium and aluminium are usually covered by a thin coating of oxide after initial reaction with oxygen in the air. Freshly produced copper is pink in colour (Figure 5.15a on p. 78). However, on exposure to air, it soon turns brown due to the formation of copper(II) oxide on the surface of the metal.

In more exposed environments, copper roofs and pipes quickly become covered in verdigris. Verdigris is green in colour (Figure 10.23) and is composed of copper salts formed on copper. The composition of verdigris varies depending on the atmospheric conditions, but includes mixed copper(II) carbonate and copper(II) hydroxide ($CuCO_3.Cu(OH)_2$).

Gold and platinum are unreactive and do not corrode, even after thousands of years.

▲ **Figure 10.23** Verdigris soon covers copper roofs in exposed environments

10.7 Alloys

The majority of the metallic substances used today are **alloys**. Alloys are mixtures of a metal with other elements. It is generally found that alloying produces a metallic substance that has more useful properties than the original pure metal it was made from. For example, the alloy brass is made from copper and zinc. The alloy is harder and more corrosion resistant than either of the metals it is made from.

Steel is a mixture of the metal iron and the non-metal carbon. Of all the alloys we use, steel is perhaps the most important. Many steel alloys have been produced; they contain iron, varying amounts of carbon, and other metals. For example, nickel and chromium are the added metals when stainless steel is produced. The chromium prevents the steel from rusting while the nickel makes it harder.

> **Key definition**
>
> **Alloys** can be harder or stronger than the pure metals and are more useful.

Alloys to order

Just as the properties of iron can be changed by alloying, so the same can be done with other useful metals. Metallurgists have designed alloys to suit a wide variety of different uses. Many thousands of alloys are now made, with the majority being 'tailor-made' to do a particular job (Figure 10.24).

Table 10.5 shows some of the more common alloys, together with some of their uses.

Many alloys are harder or stronger than the pure metals they are formed from.

a Bronze is often used in sculptures

b A polarised light micrograph of brass showing the distinct grain structure of this alloy

▲ **Figure 10.24**

In Figure 10.25, you can see that by mixing together metals which have different sized atoms to form the alloy, the metal no longer has a repeating structure. A metal is malleable and ductile because its layers are able to move over one another. In an alloy, the different sized atoms mean that the layers can no longer slide over each other, causing the alloy to have less malleability and ductility.

▲ **Figure 10.25** Alloy structure. The dark circles represent atoms of a metal; the pale circles are the larger atoms of a different metal added to make the alloy. The different size of these atoms gives the alloy different physical properties from those of the pure metal

▼ **Table 10.5** Uses of common alloys

Alloy	Composition	Use
Brass	65% copper, 35% zinc	Jewellery, machine bearings, electrical connections, door furniture
Bronze	90% copper, 10% tin	Castings, machine parts
Cupro-nickel	30% copper, 70% nickel	Turbine blades
Cupro-nickel	75% copper, 25% nickel	Coinage metal
Duralumin	95% aluminium, 4% copper, 1% magnesium, manganese and iron	Aircraft construction, bicycle parts
Magnalium	70% aluminium, 30% magnesium	Aircraft construction
Pewter	30% lead, 70% tin, a small amount of antimony	Plates, ornaments and drinking mugs
Solder	70% lead, 30% tin	Connecting electrical wiring

▶ Test yourself

15 'Many metals are more useful to us when mixed with some other elements.' Discuss this statement with respect to stainless steel.

Revision checklist

After studying Chapter 10 you should be able to:
- ✓ State the general properties of metals.
- ✓ Describe the chemical reactions of metals with acids, water/steam and oxygen.
- ✓ Give some common uses of metals and explain why a particular metal is used in that way.
- ✓ Explain what an alloy is and why its properties are different to the pure metals it contains.
- ✓ State and explain the order of the reactivity series and use it to predict the chemical reactions of the metals.
- ✓ State that unreactive metals occur uncombined (not as compounds) and describe how the method used to extract reactive metals from their ores depends on their position in the reactivity series.

- ✓ Describe displacement experiments which could be carried out to find the order of reactivity of metals.
- ✓ Explain why aluminium is so unreactive.
- ✓ Describe how iron is extracted from its ore, hematite.
- ✓ Describe methods for preventing the corrosion of metals.
- ✓ Explain how zinc is able to sacrificially protect iron.
- ✓ Explain why alloys are often harder and stronger than the metals they are composed of.

Exam-style questions

1 Use the following list of metals to answer the questions **a** to **i**: iron, calcium, potassium, gold, aluminium, magnesium, sodium, zinc, platinum.

   a Identify the metals that are found native. [1]

   b Identify which of the metals is found in nature as the ore:

      i rock salt [1]

      ii bauxite. [1]

   c Identify the metal that has a carbonate found in nature called chalk. [1]

   d Identify which of the metals will not react with oxygen to form an oxide. [1]

   e Identify which of the metals will react violently with cold water. [1]

   f Choose one of the metals in your answer to **e** and give a balanced chemical equation for the reaction which takes place. [1]

   g Identify which of the metals has a protective oxide coating on its surface. [1]

   h Identify which of the metals reacts very slowly with cold water but extremely vigorously with steam. [1]

   i Identify which of the metals is used to galvanise iron. [1]

2

Copper

↓

heat in air (oxygen)

↓

Black solid **A**

↓

heat with powdered magnesium

↓

Brown–pink solid **B** +

white powder **C**

↓

dilute hydrochloric acid

↓

Brown–pink solid **B** +

colourless solution **D** + water

↓

filter and evaporate filtrate to dryness

↓

Solid **D**

↓

electrolysis of molten **D**

↓

Silvery metal **E** + green gas **F**

   a Give the name and formulae of the substances **A** to **F**. [6]

   b Give balanced chemical equations for the reactions in which:

      i black solid **A** was formed [1]

      ii white powder **C** and brown–pink solid **B** were formed [1]

      iii colourless solution **D** was formed. [1]

   c 'The reaction between black solid **A** and magnesium is a redox reaction.' With reference to this reaction, explain what you understand by this statement. [3]

   d Write anode and cathode reactions for the processes which take place during the electrolysis of molten **D**. [2]

   e Suggest a use for:

      i brown–pink solid **B** [1]

      ii silvery metal **E** [1]

      iii green gas **F**. [1]

3 Explain the following:

   a Metals, such as gold and silver, occur native in the Earth's crust. [2]

   b The parts of shipwrecks made of iron rust more slowly in deep water. [2]

   c Zinc bars are attached to the structure of oil rigs to prevent them from rusting. [2]

   d Copper roofs quickly become covered with a green coating when exposed to the atmosphere. [2]

   e Recycling metals can save money. [2]

4 Iron is extracted from its ores hematite and magnetite. Usually it is extracted from hematite (iron(III) oxide). The ore is mixed with limestone and coke, and reduced to the metal in a blast furnace. The following is a brief outline of the reactions involved.

   $$\text{coke} + \text{oxygen} \rightarrow \text{gas } \mathbf{X}$$

   $$\text{gas } \mathbf{X} + \text{coke} \rightarrow \text{gas } \mathbf{Y}$$

   $$\text{iron(III) oxide} + \text{gas } \mathbf{Y} \rightarrow \text{iron} + \text{gas } \mathbf{X}$$

   a Name the gases **X** and **Y**. [2]

   b Give a chemical test to identify gas **X**. [2]

   c Write balanced chemical equations for the reactions shown above. [3]

d The added limestone is involved in the following reactions:

limestone → calcium oxide + gas **X**

calcium oxide + silicon(IV) oxide → slag

  i Give the chemical names for limestone and slag. [2]

> **ii** Write balanced chemical equations for the reactions shown above. [2]

  iii Explain why the reaction between calcium oxide and silicon(IV) oxide is called an acid–base reaction. [2]

  iv Describe what happens to the liquid iron and slag when they reach the bottom of the furnace. [2]

e Suggest why the furnace used in the extraction of iron is called a blast furnace. [1]

5 Zinc can be reacted with steam using the apparatus shown. When gas **A** is collected, mixed with air and ignited, it gives a small pop. A white solid **B** remains in the test tube when the reaction has stopped and the apparatus cooled down.

a Give the name and formula of gas **A**. [2]

b **i** Identify the product formed when gas **A** burns in air. [1]

  ii Give a balanced chemical equation for this reaction. [1]

c **i** Identify white solid **B**. [1]

  ii Give a balanced chemical equation to represent the reaction between zinc and steam. [1]

d Identify another metal which could be safely used to replace zinc and produce another sample of gas **A**. [1]

e When zinc reacts with dilute hydrochloric acid, gas **A** is produced again. Give a balanced chemical equation to represent this reaction and name the other product of this reaction. [2]

6 Students set up the experiment below to find out the conditions which were needed for rusting to occur.

a Give the purpose of:

  i the anhydrous calcium chloride in tube D [1]

  ii boiling the water in tube A [1]

  iii the layer of oil in tube A. [1]

b **i** Identify which of the tubes contain air. [1]

  ii Identify which of the tubes contain water. [1]

  iii Identify which tube(s) contain air and water. [1]

c Identify the tube(s) in which the nails will not rust. [1]

Chemistry of the environment

In this chapter, you will learn about the most plentiful liquid on this planet, **water**, and the most important gaseous mixture we know, the **air** (or atmosphere). The importance of water and air is not in doubt. Without either of these, life would not be possible on Earth! Also, we consider the problems with the pollution we have created worldwide, including excessive use of fertilisers, fossil fuels and plastics, and the greenhouse effect and the associated global warming. You will also learn about some possible solutions to the problems. By the end of this chapter, you will be able to provide a possible answer to this question:

How can we guard our planet against the problems we have and those we continue to create?

11.1 Water

Water is the commonest compound on this planet. More than 70% of the Earth's surface is covered with sea, and the land masses are dotted with rivers and lakes (Figure 11.1a). It is vital to our existence and survival because it is one of the main constituents in all living organisms. For example, your bones contain 72% water, your kidneys are about 82% water and your blood is about 90% water (Figure 11.1b).

a Millions of tonnes of water pass over this waterfall every day

b Your blood contains a lot of water

▲ **Figure 11.1**

Going further

Water circulates around the Earth via the water cycle (Figure 11.2). The driving force for the water cycle is the Sun.

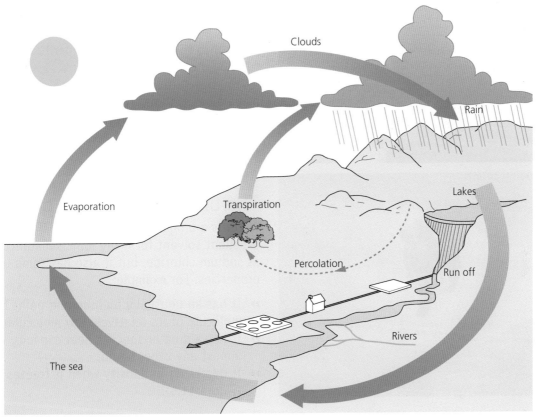

▲ **Figure 11.2** The water cycle

● Heat from the Sun causes evaporation from oceans, seas and lakes. Water vapour is also formed from the evaporation of water from leaves (transpiration), through respiration and through combustion. The water vapour rises and cools, and condenses to form tiny droplets of water. These droplets form clouds.
● The clouds are moved along by air currents. As they cool, the tiny droplets join to form larger droplets, which fall as rain when they reach a certain size.

● The water that falls as rain runs into streams and rivers and then on into lakes, seas and oceans.

Some water is diverted from rivers into reservoirs and purified (p. 173). This water is then used in industry and the home. After use it is dirty and has to be treated at the sewage works before being returned to a river or the sea.

Water has many other important uses besides sustaining life. These include:

In the home:

» cooking
» cleaning
» drinking.

In industry:

» as a solvent
» as a coolant
» for cleaning
» as a chemical reactant.

Pure water is a neutral, colourless liquid which (at 1 atmosphere pressure) boils at 100°C and freezes at 0°C (Figure 11.3). If you try boiling tap water, you will find that it does not boil at exactly 100°C. This is because it is a very good solvent and dissolves many substances. It is the presence of these impurities that causes a change in the boiling point compared to pure water. Distilled water is therefore used in many experiments in the laboratory because it is very pure and does not contain many of these impurities, which may affect the outcome of the experiments.

▲ **Figure 11.3** Liquid water boils at 100°C and freezes to form ice at 0°C

You can find out whether a colourless liquid contains water by adding the unknown liquid to anhydrous copper(II) sulfate. If this changes from white to blue, then the liquid contains water (Figure 11.4a).

Another test is to dip blue cobalt(II) chloride paper into the liquid. If the paper turns pink, then the liquid contains water (Figure 11.4b).

You have already seen in Chapter 5 that water may be electrolysed (when acidified with a little dilute sulfuric acid). When this is done, the ratio of the volume of the gas produced at the cathode to that produced at the anode is 2 : 1. This is what you might expect, since the formula of water is H_2O!

The unique properties of water

Water is a unique substance. Not only is it an excellent solvent for many ionic substances, such as sodium chloride, but it also has some unusual properties. For example:

» It has an unusually high boiling point for a molecule of its relatively low molecular mass.
» It has a greater specific heat capacity than almost any other liquid.
» It decreases in density when it freezes (Figure 11.5).

a Anhydrous copper(II) sulfate goes blue when water is added to it

▲ **Figure 11.4** Tests for the presence of water

b Cobalt(II) chloride paper turns pink when water is dropped on to it

▲ **Figure 11.5** When water freezes, its density falls; this is why icebergs float

⮕ Going further

These unusual properties can be explained by **hydrogen bonding**. This is a weak intermolecular force (bond) which occurs between water molecules because the bonds within the molecules are polar. A polar bond is one which contains, for example, oxygen attached to hydrogen by a covalent bond. The shared electrons in the bond are drawn towards the oxygen atom to create a small negative charge on the oxygen atom ($\delta-$). There is then an equally small positive charge on the hydrogen atom ($\delta+$). The water molecules then attract one another as shown in Figure 11.6. In the case of water, this attraction is called a hydrogen bond. It is a much weaker bond than a covalent bond.

▲ **Figure 11.6** The unusual properties of water can be explained by hydrogen bonding

Water pollution and treatment

An adequate supply of fresh and safe water is essential to the health and well-being of the world's population. Across the planet, biological and chemical pollutants are affecting the quality of our water. Lack of availability of fresh water leads to waterborne diseases, such as cholera and typhoid, and to diarrhoea, which is one of the biggest killers across the world.

Agriculture needs a water supply in order to irrigate crops, especially in areas of the world with hot climates. The production of more and more crops for the ever-increasing population is essential.

> **Key definition**
>
> **Pollution** is a modification that takes place to the environment, water or air, caused by human influence such as releasing substances into our water or air.

Water is very good at dissolving substances and so it is very unusual to find really pure water on this planet. As water falls through the atmosphere, on to and then through the surface of the Earth, it dissolves a tremendous variety of substances including:

» a variety of gases from the air such as carbon dioxide and oxygen. However, oxygen is beneficial to aquatic life and allows it to thrive in our rivers and oceans. Carbon dioxide creates a low-level natural acidity in the water, but this acidity is increasing, due to the increase in the amount of carbon dioxide released into the atmosphere by vehicles and industry.

» nitrates and phosphates from agricultural waste and detergents. Chemical fertilisers, washed off surrounding land, add nitrate ions (NO_3^-) and phosphate ions (PO_4^{3-}) to the water. This is due to the use of artificial fertilisers, such as ammonium nitrate and ammonium phosphate (see below), as well as some pesticides. Detergents used in the home and in industry contain phosphates. The nitrates and phosphates encourage the growth of algae which eventually die and decay, removing oxygen from the water (deoxygenation). They can also disrupt sensitive ecosystems.

» metal compounds from industrial waste water, which may contain harmful chemicals, such as cadmium and mercury, which are **toxic**. It should be noted however, that not all metal compounds are toxic. Some metal compounds that are present are also beneficial as they provide essential minerals. This is especially true of calcium-containing compounds. The calcium present in water is necessary for healthy growth of bones and teeth.

» human waste from sewage. Sewage contains harmful microbes which cause disease.

» insoluble impurities such as oil and plastic waste (Figure 11.7). Plastic waste is polluting our streams and rivers as well as oceans. It is detrimental to all aquatic life as well as human life on this planet. This issue will be discussed in more detail in Chapter 12.

▲ **Figure 11.7** A badly polluted river

All these artificial and natural impurities must be removed from the water before it can be used. Recent regulations in many countries have imposed strict guidelines on the amounts of various substances allowed in drinking water.

▲ **Figure 11.8** This lake is used as a source of drinking water

A lot of drinking water is obtained from lakes and rivers where the pollution levels are low (Figure 11.8). Undesirable materials are removed from water by the process of water treatment, which involves both filtration and chlorination and is summarised in Figure 11.9.

1 Impure water is passed through screens to filter out floating debris.
2 Aluminium sulfate is added to coagulate small particles of clay so that they form larger clumps, which settle more rapidly.

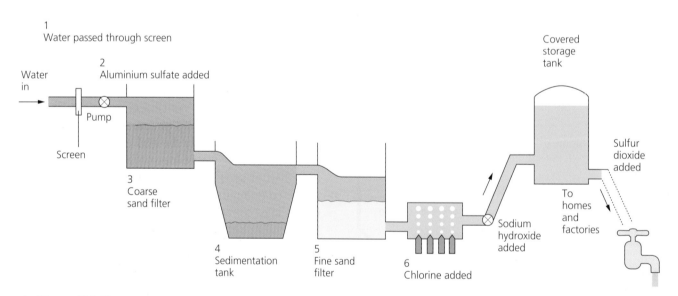

▲ **Figure 11.9** The processes involved in water treatment

3 Filtration through coarse sand traps larger, insoluble particles. The sand also contains specially grown microbes which remove some of the bacteria.

4 A sedimentation tank has chemicals known as flocculants, for example, aluminium sulfate, added to it to make the smaller particles (which remain in the water as colloidal clay) stick together and sink to the bottom of the tank.

5 These particles are removed by further filtration through fine sand. This is followed by carbon slurry filters which are there to remove unwanted tastes and odours, and a lime slurry is used to adjust the acidity.

6 Finally, a little chlorine gas is added, which sterilises the water and kills any remaining bacteria. Excess chlorine can be removed by the addition of sulfur dioxide gas. The addition of chlorine gas makes the water more acidic and so appropriate amounts of sodium hydroxide solution are added. Fluoride is sometimes added to water if there is insufficient occurring naturally, as it helps to prevent tooth decay.

11.2 Artificial fertilisers

Some of the ammonia produced by the Haber process (see Chapter 7, p. 109) is used to produce nitric acid. If ammonia is then reacted with the nitric acid, ammonium nitrate is produced. This gives us the basic reaction for the production of many artificial **fertilisers**.

> **Key definition**
>
> **Fertiliser** is a chemical put onto soil to replace lost mineral salts and so make plants grow more healthily. These include ammonium salts, such as ammonium nitrate, which is one of the most commonly used fertilisers.

$$\text{ammonia} + \text{nitric acid} \rightarrow \text{ammonium nitrate}$$

$$NH_3(g) + HNO_3(aq) \rightarrow NH_4NO_3(aq)$$

Ammonium nitrate (Nitram®) is probably the most widely used nitrogenous fertiliser. The use of artificial fertilisers is essential if farmers are to produce sufficient crops to feed the ever-increasing world population. Crops remove nutrients from the soil as they grow; these include nitrogen, phosphorus and potassium. Artificial fertilisers are added to the soil to replace these nutrients

and others, such as calcium, magnesium, sodium, sulfur, copper and iron. Examples of nitrogenous fertilisers (those which contain nitrogen) are shown in Table 11.1.

▼ **Table 11.1** Some nitrogenous fertilisers

Fertiliser	Formula
Ammonium nitrate	NH_4NO_3
Ammonium phosphate	$(NH_4)_3PO_4$
Ammonium sulfate	$(NH_4)_2SO_4$
Urea	$CO(NH_2)_2$

Artificial fertilisers can also create fertile land from areas unable to support crop growth. The fertilisers which add the three main nutrients (N, P and K) are called NPK fertilisers. They contain ammonium nitrate (NH_4NO_3), ammonium phosphate (($NH_4)_3PO_4$) and potassium chloride (KCl) in varying proportions (Figure 11.10). Fertilisers have an important role in the nitrogen cycle.

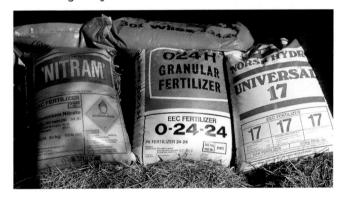

a Different fertilisers contain differing amounts of the elements nitrogen, phosphorus and potassium

b The different NPK elements are responsible for the healthy growth of plants in different ways

▲ **Figure 11.10**

Going further

The vital importance of nitrogen to both plants and animals can be summarised by the nitrogen cycle (Figure 11.12).

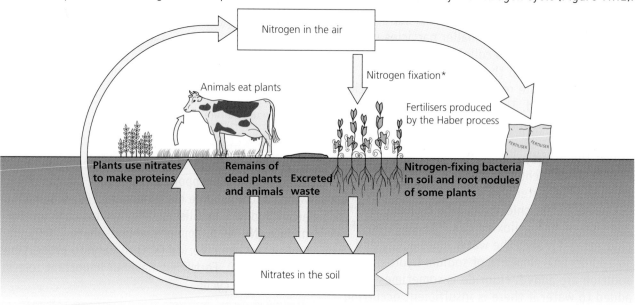

▲ **Figure 11.12** The nitrogen cycle
*Nitrogen fixation is the direct use of atmospheric nitrogen in the formation of important compounds of nitrogen. Bacteria present in the root nodules of certain plants are able to take nitrogen directly from the atmosphere to form essential protein molecules.

If farm crops are harvested from the land rather than left to decay, the soil becomes deficient in this important element. The nitrogen is removed in the harvested crops rather than remaining as the plants decay. In addition, nitrates can be washed from the soil by the action of rain (leaching). For the soil to remain fertile for the next crop, the nitrates need to be replaced. The natural process is by decay or by the action of lightning on atmospheric nitrogen. Without the decay, however, the latter process is not efficient enough to produce nitrates on the scale required.

Farmers often need to add substances containing these nitrates. Such substances include farmyard manure and artificial fertilisers. One of the most commonly used artificial fertilisers is ammonium nitrate which, as you saw earlier, is made from ammonia gas and nitric acid, both nitrogen-containing compounds.

▲ **Figure 11.11** Fertilisers have been used to help create some of the best fruit and vegetables on sale

? **Worked example**

Using relative atomic masses and the formula of a fertiliser, it is possible to calculate the percentage of each of the essential elements in a fertiliser.

What is the percentage of nitrogen in the fertiliser ammonium sulfate $(NH_4)_2SO_4$?

$(A_r$: H = 1; N = 14; O = 16; S = 32)

The formula mass of ammonium sulfate $(NH_4)_2SO_4$ is:

$(2 \times 14) + 2(4 \times 1) + 32 + (4 \times 16) = 132$

The % nitrogen in ammonium sulfate is:

$((2 \times 14)/132) \times 100 = 21.2\%$

Problems with fertilisers

If artificial fertilisers of all kinds are not used correctly, problems can arise. If too much fertiliser is applied to the land, rain washes the fertiliser off the land and into rivers and streams. This is known as leaching. This leaching leads to **eutrophication**: the process that occurs when fertiliser is leached, causing algae to multiply rapidly and causing the water to turn green. As the algae die and decay, oxygen is removed from the water, leaving insufficient amounts for fish and other organisms to survive (Figure 11.13). In extreme cases, no normal aquatic life can survive. There are also worries about the effect of agricultural fertilisers, especially nitrates such as ammonium nitrate, on the public water supply.

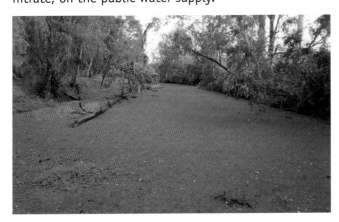

▲ **Figure 11.13** Over-use of fertilisers has led to eutrophication

Test yourself

1 Calculate the percentage of nitrogen in each of the three fertilisers ammonium nitrate (NH_4NO_3), ammonium phosphate (($NH_4)_3PO_4$) and urea ($CO(NH_2)_2$). (A_r: H = 1; C = 12; N = 14; O = 16; P = 31)

2 Write down a method that you could carry out in a school laboratory to prepare a sample of ammonium sulfate fertiliser.

3 Why is it important to have nitrogen in fertilisers?

11.3 The air

➡ Going further

The gases in the air are held in an envelope around the Earth by its gravity. The atmosphere is approximately 100 km thick (Figure 11.14),

and about 75% of the mass of the atmosphere is found in the layer nearest the Earth called the troposphere (Figure 11.15). Beyond this layer, the atmosphere reaches into space but becomes extremely thin. Nearly all atmospheric water vapour (or moisture) is found in the troposphere, which also contains the liquid water in the oceans, rivers and lakes.

▲ **Figure 11.14** The lighter blue shows the extent of the atmosphere around the Earth – approximately 100 km

▲ **Figure 11.15** The Earth's atmosphere

The composition of the atmosphere

If a sample of dry, unpolluted air were taken from any location in the troposphere and analysed, the composition by volume of the sample would be similar to that shown in Table 11.2.

▼ **Table 11.2** Composition of the atmosphere

Component	%
Nitrogen	78.08
Oxygen	20.95
Argon	0.93
Carbon dioxide	0.04
Neon	0.002
Helium	0.000 5
Krypton	0.000 1
Xenon plus tiny amounts of other gases	0.000 01

Note that air contained approximately 0.03% by volume of carbon dioxide for many years. This value has remained almost constant for a long period of time and is maintained via the processes found in the carbon cycle (Figure 11.16).

However, scientists have recently detected an increase in the amount of carbon dioxide in the atmosphere to approximately 0.04%.

> **Key definition**
>
> Clean, dry air is approximately 78% nitrogen, 21% oxygen and the remainder is a mixture of noble gases and carbon dioxide.

Sources of carbon dioxide in the air

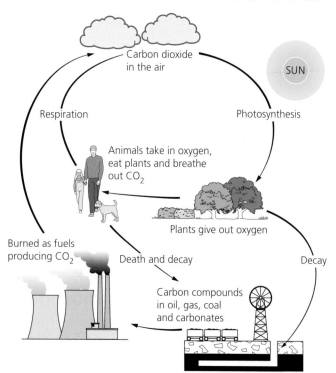

▲ **Figure 11.16** The carbon cycle

Figure 11.16 shows how carbon dioxide is produced by the combustion of fossil fuels such as methane found in natural gas. It is also produced by all living organisms. Animals take in oxygen and breathe out carbon dioxide.

Carbon dioxide is taken in by plants through their leaves and used together with water, taken in through their roots, to synthesise glucose (a sugar). This is the process of **photosynthesis**.

It takes place only in sunlight and only in green leaves, as they contain chlorophyll (the green pigment) which catalyses the process.

$$\text{carbon dioxide} + \text{water} \xrightarrow[\text{chlorophyll}]{\text{sunlight}} \text{glucose} + \text{oxygen}$$

$$6CO_2 + 6H_2O \rightarrow C_6H_{12}O_6 + 6O_2$$

The carbon cycle has continued in this manner for millions of years. However, scientists have detected an imbalance in the carbon cycle due to the increase in the amount of carbon dioxide produced through burning fossil fuels and the deforestation of large areas of tropical rainforest.

The Earth's climate is affected by the levels of carbon dioxide (and water vapour) in the atmosphere. If the amount of carbon dioxide, in particular, builds up in the air, it is thought that the average temperature of the Earth will rise, causing global warming. This effect is thought by scientists to be caused by the greenhouse effect. This important topic will be discussed further later in this chapter on p. 183.

> **Test yourself**
>
> 4 Draw a pie chart to show the data given in Table 11.2.
> 5 Is air a compound or a mixture? Explain your answer.
> 6 Which of the following statements about the air are true and which are false?
>   a Nitrogen constitutes nearly 79% of the atmosphere.
>   b Air is a mixture of elements.
>   c Carbon dioxide is not a constituent of the air.
>   d Neon is present in the air.

11.4 Atmospheric pollution

The two major resources considered in this chapter, water and air, are essential to our way of life and our very existence. Water and air make up the environment of a living organism. The environment is everything in the surroundings of an organism that could possibly influence it. Humans continually pollute these resources. We now look at the effects of the various sources of pollution of the air and at the methods used to control or eliminate them. For a discussion of water pollution, see p. 173.

Air pollution is all around us. Concentrations of gases in the atmosphere, such as carbon monoxide, sulfur dioxide and nitrogen oxides, are increasing with the increasing population. As the population increases, there is a consequent increase in the need for energy, industries and motor vehicles. These gases are produced primarily from the combustion of the fossil fuels, coal, oil and gas, but they are also produced by the smoking of cigarettes.

Motor vehicles are responsible for much of the air pollution in large towns and cities. They produce five particularly harmful pollutants, as shown in Table 11.3.

▼ **Table 11.3** Pollution caused by motor vehicles

Pollutant	Caused by	Problems caused	Can be reduced by
Carbon dioxide, CO_2	The complete combustion of carbon-containing fuels	Higher levels of carbon dioxide lead to increased global warming, which leads to climate change	Burning fewer fossil fuels
Carbon monoxide, CO	The incomplete combustion of carbon-containing fuels	Toxic gas	Fitting catalytic converters to remove CO, but more CO_2 is produced. It should be noted that lean-burn engines also produce less CO.
Particulates, C	The incomplete combustion of carbon-containing fuels	Increased risk of respiratory problems and cancer	Adding oxygenates such as ethanol to petrol to reduce the particulates produced
Methane, CH_4	The decomposition of vegetation and waste gases from digestion in animals	Higher levels of methane lead to increased global warming, which leads to climate change	Encouraging composting or incineration. Capturing the methane produced and using it as a fuel or to generate electricity when burned. Reducing the number of farm animals by encouraging people to eat less meat
Nitrogen oxides, NO_x	Car engines	Acid rain, photochemical smog and respiratory problems	Fitting catalytic converters to remove nitrogen oxides. However, in doing so more carbon dioxide is produced.
Sulfur dioxide, SO_2	The combustion of fossil fuels which contain sulfur compounds	Acid rain	Removing sulfur from petrol to produce low-sulfur petrol. Using flue gas desulfurisation units (see p. 182) at coal-burning power stations

Catalytic converters

In recent years, regulations have been introduced worldwide which state that all new cars have to be fitted with **catalytic converters** as part of their exhaust system (Figure 11.17). Car exhaust fumes contain pollutant gases, such as carbon monoxide (CO), formed from the incomplete combustion of hydrocarbons(C_xH_y) in the fuel, and nitrogen(II) oxide (NO) formed by the reaction of nitrogen gas and oxygen gas from the air. The following reactions proceed of their own accord but very slowly under the conditions inside an exhaust.

> nitrogen(II) + carbon → nitrogen + carbon
> oxide   monoxide   dioxide
> $$2NO(g) + 2CO(g) \rightarrow N_2(g) + 2CO_2(g)$$

The catalyst in the catalytic converter speeds up these reactions considerably. The catalyst is made from platinum or a similar platinum-like metal such as palladium or rhodium. In these reactions, the pollutants are converted to carbon dioxide and nitrogen, which are naturally present in the air. The removal of oxides of nitrogen is important because they cause respiratory disease. They are also involved in the production of photochemical **smogs** (Figure 11.18) which occur worldwide in major cities, especially in the summer. Photochemical smog is the most widely known and perhaps most serious

air pollutant. It is formed in the atmosphere by the reaction between gaseous pollutants, nitrogen oxides and hydrocarbons.

a A catalytic converter

b A section through a catalytic converter

▲ **Figure 11.17**

▲ **Figure 11.18** The haze is caused by a photochemical smog, which is due to pollution caused mainly by cars without catalyst exhaust systems

Going further

The incomplete combustion of hydrocarbons in the petrol and the evaporation of hydrocarbons from petrol tanks both contribute to the formation of photochemical smog. This can cause severe respiratory problems. However adding oxygenates, such as ethanol, to petrol reduces its production. Catalytic converters also help remove unburned hydrocarbons but in doing so more carbon dioxide is produced!

Test yourself

7   Catalytic converters remove harmful gases CO and NO. They do this by converting them into less harmful chemicals. Give the names and formulas of the chemicals they are converted to.

A further method of regulating pollutant gases from vehicles is to remove petrol-burning engines and replace them with an efficient electric motor. The research and development of these motors is moving forward at a tremendous pace.

Internationally there has been great concern globally with the increase of **particulates** in the air (Figure 11.19). 'Particulates' is a general term used to describe very small particles in the atmosphere, such as certain types of smoke emitted from diesel engines, as well as dust.

 Going further

These particulates have been associated with a variety of lung complaints in both adults and children. It has been found, however, that adding oxygenates, such as ethanol, to fuels reduces the amount of particulates produced by vehicles. The oxygenates provide extra oxygen to the burning process in the engine and so reduce the amount of incomplete combustion taking place.

▲ **Figure 11.19** Particulates produced by diesel engines are a real health problem worldwide

In addition to the fuels used in motor vehicles producing quantities of sulfur dioxide when they combust in engines, heavy industry (Figure 11.20) and power stations are also major sources of sulfur dioxide, formed by the combustion of coal, oil and gas, which also contain small amounts of sulfur.

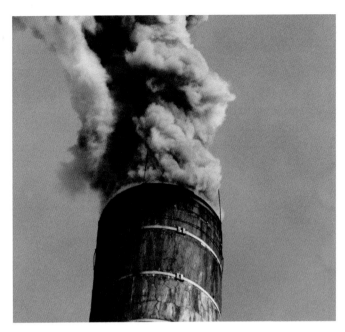

▲ **Figure 11.20** Sulfur dioxide is a major pollutant produced by industry

Rainwater is naturally acidic since it dissolves carbon dioxide gas from the atmosphere as it falls. Natural rainwater has a pH of about 5.7. However, the acidity is enhanced by sulfur dioxide gas dissolving in rainwater to form the weak acid, sulfurous acid (H_2SO_3).

A further reaction occurs in which the sulfurous acid is oxidised to sulfuric acid (H_2SO_4). Solutions of these acids are the principal contributors to acid rain. It should be noted that the reduction of nitrogen(IV) oxide by the use of catalytic exhaust systems is important since the NO_2 produced in the atmosphere dissolves in rainwater and produces, eventually, nitric acid which adds to acid rain. In recent years, especially in central Europe, the pH of rainwater has fallen to between pH 3 and pH 4.8. This increase in acidity has led to extensive damage to forests (Figure 11.21), lakes and marine life.

▲ **Figure 11.21** This forest has been devastated by acid rain

Practical skills

Acid rain

For safe experiments/demonstrations which are related to this chapter, please refer to the *Cambridge IGCSE Chemistry Practical Skills Workbook*, which is also part of this series.

Safety

- Eye protection must be worn.

Experiment A

A student tests the rainwater collected from the school playing field using universal indicator solution. They then tested rainwater from three other places, A, B and C.

1 What apparatus would be required for this experiment?
   The results of the tests are shown in the table below.

Sample	Colour of universal indicator solution	pH of rain water
Playing fields	Lime green	6
A	Orange	3
B	Yellow	5
C	Lime green	6

2 Which was the most acidic water?
3 Which was the least acidic water?

Experiment B

The student was then given some powdered calcium carbonate to add to each sample. The results are shown in the table below. The student was asked to collect samples of any gas produced and test any gas produced using limewater.

Sample	Observation upon adding calcium carbonate powder	Observation after adding limewater
Playing Fields	Small amount of bubbles but not sufficient to test	No result
A	Lots of bubbles of a gas produced	Limewater turns milky white
B	Small amount of bubbles but not sufficient to test	No result
C	Small amount of bubbles but not sufficient to test	No result

4 a Which gas was being tested for?
   b How could a sample of the white solid be obtained after the test was carried out on **A**?
5 Describe, with the aid of a simple diagram, how the student could collect samples of any gas produced.
6 Did these results confirm the results obtained in experiment A?
   Explain your answer.
7 Write a conclusion to experiments A and B.

Recently, units called flue gas desulfurisation (FGD) units have been fitted to some power stations throughout the world to prevent the emission of sulfur dioxide gas. Here, the sulfur dioxide gas is removed from the waste gases by passing them through calcium oxide slurry. This not only removes the sulfur dioxide but also creates calcium sulfite, which then oxidises to sulfate, which can be sold to produce plasterboard (Figure 11.22). The FGD units are very expensive and therefore the sale of the calcium sulfate is an important economic part of the process.

▲ **Figure 11.22** This plasterboard is made using calcium sulfate from an FGD plant

Global warming

Scientists know that certain gases in the atmosphere trap the Sun's heat in our atmosphere and this makes the Earth warmer. These gases are known as greenhouse gases and cause the greenhouse effect. This increase in the Earth's temperature is known as global warming.

It is well known by scientists that, for example, higher levels of carbon dioxide in the atmosphere lead to increased global warming, which leads to climate change.

How does the greenhouse effect affect the average temperature of the Earth?

Some energy from the Sun is absorbed by the Earth and its atmosphere. The remainder is reflected back into space. The energy that is absorbed helps to heat up the Earth. The Earth radiates some heat energy back into space but the greenhouse gases, including carbon dioxide, prevent it from escaping. This effect is similar to that observed in a greenhouse where sunlight (visible/ultraviolet radiation) enters through the glass panes but heat (infrared radiation) has difficulty escaping through the glass (Figure 11.23).

Carbon dioxide is not the only greenhouse gas. Other gases also contribute to the greenhouse effect. One of these is methane (see Chapter 12, p. 189), which is produced from agriculture and released from landfill sites as well as rice fields. The amount of carbon dioxide in the atmosphere is increasing due to:

» major deforestation taking place in several countries of the world. Trees act as the lungs of the planet. During photosynthesis they absorb carbon dioxide and release oxygen.
» pollution of our rivers and oceans. The more pollutants that are dissolving in these systems, the less carbon dioxide that can dissolve in the water.
» a large increase in the amounts of carbon dioxide released by the continued burning of large quantities of fossil fuels by industry and transport.

The long-term effect of the higher temperatures of the greenhouse effect and the subsequent global warming will be the continued gradual melting of ice caps and consequent flooding in low-lying areas of the Earth. There will also be further changes in the weather patterns which will further affect agriculture worldwide.

There is an additional effect when the incomplete combustion of diesel takes place. Carbon particles or particulates are produced. These black particles are blown by the prevailing winds and some end up falling on snow and ice fields. This reduces the reflectivity of the snow, which causes a further increase in the melting of the snow and ice in mountainous areas as well as the northern and southern pole areas of the world.

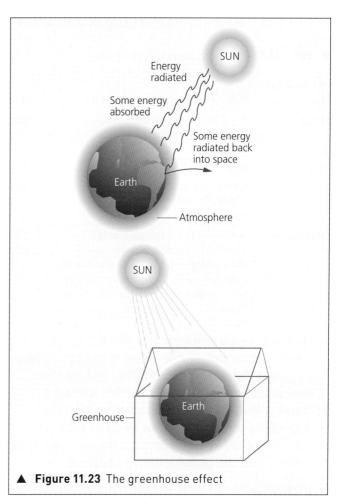

▲ **Figure 11.23** The greenhouse effect

These problems have been recognised by nations worldwide. Recent agreements under the Kyoto Protocol, as well as the Paris Accord, between nations mean that there will be some reduction in the amount of carbon dioxide (and other greenhouse gases) produced over the next few years. However, there is still a long way to go before we no longer increase greenhouse gases in the atmosphere.

How can we reduce the effects of global warming?

There are several actions that can be taken to reduce the effects of global warming:

» Plant more trees and reduce the amount of deforestation taking place worldwide. This will cause more carbon dioxide to be absorbed from the atmosphere through photosynthesis.
» Burn fewer fossil fuels by using alternative fuels in order to reduce the amount of carbon dioxide entering the atmosphere. You will see in Table 11.4 that there are alternatives that might improve matters. However, we must take into account whether or not the alternative fuel is going to use up other resources in a way that enables them to continue in the future rather than destroying them. In other words, is the alternative sustainable? Sustainability is very important. In the case of fossil fuels, they are not renewable or sustainable. When they are gone they are gone! Also we must consider the advantages and disadvantages of the alternative fuel. Table 11.4 shows not only whether those alternatives are sustainable but also their perceived advantages and disadvantages.

▼ **Table 11.4** Alternative fuels to fossil fuels

Alternative fuel	Is the alternative sustainable?	Advantages	Disadvantages
Biodiesel	Yes it is! It is made from waste plant material and animal oils and fats.	Produces less CO, C_xH_y, SO_2 and particulates than diesel fuel.	NO emissions are higher than from standard diesel fuel.
Ethanol	This is open to debate since large amounts of energy and land are needed to cultivate sugar cane for fermentation.	Less CO, SO_2 and NO_x are produced than from petrol. Replanting sugar cane creates a cycle as it absorbs CO_2 from the atmosphere.	It is very flammable.
Hydrogen	Hydrogen is sustainable only if the electricity needed to produce it, from the electrolysis of acidified water, is from a renewable resource such as solar power or wind. The car shown in Figure 11.24 is powered by hydrogen.	Water is the only product of production. Hence there is no pollution.	It is very, very flammable. Also a high-pressure fuel tank is needed to store it as a liquid.

» Be more energy efficient. This will reduce the amount of carbon dioxide entering the atmosphere. In homes, for example, this means making them well insulated and turning off lights when not needed. We also need to consider whether the journey we are about to take by car or plane is essential. Only by being conscientious will we make inroads into the global warming situation.

» Increase the use of renewable energy sources such as solar, wind and tidal.
» Significantly reduce livestock farming to cut down the amount of methane released into the atmosphere.

▲ **Figure 11.24** A hydrogen-powered vehicle

➜ Going further

Some uses of the gases from the air

Our everyday life would not be the same without the gases we obtain from the air. For example:

● Oxygen is used in hospitals by patients with breathing difficulties as well as in the manufacture of steel.
● Nitrogen is used in the manufacture of fertilisers (see p. 175) as well as in food packaging to keep food fresh.
● Inert gases, such as argon, are used to fill the space between the panes of modern double-glazing units, while neon is used in advertising signs.

Test yourself

8 In 1960, the percentage of carbon dioxide in the atmosphere was close to 0.03% and in 2010 it was close to 0.04%. Suggest an explanation for this increase.

9 Write a balanced chemical equation to represent the reaction which takes place between sulfur dioxide and calcium hydroxide slurry in the FGD unit of a power station.

10 Write down one problem that can be caused by each of these air pollutants:
   a nitrogen dioxide
   b particulates.

Revision checklist

After studying Chapter 11 you should be able to:
✔ Name the gases in the atmosphere and state the approximate percentage proportions of each gas.
✔ Describe chemical tests for the presence of water.
✔ Describe how to test for the purity of water.
✔ Explain why distilled water is used in practical chemistry rather than tap water.
✔ State that ammonium salts and nitrates can be used as fertilisers, including ammonium phosphate and potassium nitrate.
✔ Describe the use of NPK fertilisers to provide improved plant growth and provide larger yields of crops to feed the world's growing population.
✔ State the composition of clean, dry air.
✔ State the source of these air pollutants: carbon dioxide, carbon monoxide, methane, nitrogen oxides, sulfur dioxide and particulates.

✔ State the adverse effect of these air pollutants.

✔ Describe how the greenhouse gases, such as carbon dioxide and methane, cause global warming, which leads to climate change.
✔ Explain how oxides of nitrogen form in car engines and describe their removal by catalytic converters.

✔ State and explain strategies to reduce the effects of these various environmental issues.
✔ Describe photosynthesis as the reaction between carbon dioxide and water to produce glucose in the presence of chlorophyll and using energy from light .

Exam-style questions

1 The apparatus shown below was used to estimate the proportion of oxygen in the atmosphere.

100 cm³ of air is pushed out of this syringe

Across hot copper

Into this syringe

Heat

Air is passed backwards and forwards across the copper, which turns black

The volume of air gets smaller and smaller

Eventually the volume of air stops getting smaller

The volume of remaining air is measured

Heat

A volume of dry air (200 cm³) was passed backwards and forwards over heated copper until no further change in volume took place. The apparatus was then allowed to cool down to room temperature and the final volume reading was then taken. Some typical results are shown below.

Volume of gas before = 200 cm³

Volume of gas after = 157 cm³

During the experiment, the copper slowly turned black.

a Explain why the apparatus was allowed to cool back to room temperature before the final volume reading was taken. [2]

b Using the information given above, calculate the percentage volume reduction which has taken place. [3]

c Explain briefly why there is a change in volume. [2]

d Identify which observation given above supports your explanation in c.

e Give a balanced chemical equation for any reaction which has occurred. [4]

f Give the name of the main residual gas at the end of the experiment. [1]

g Would you expect the copper to have increased or decreased in mass during the experiment? Explain your answer. [2]

2 a Oxygen has an atomic number of 8 and mass number 16. Give the electronic configuration of the oxygen atom. [1]

b How many electrons, neutrons and protons are there in the oxygen atom? [3]

c Oxygen molecules are diatomic. Explain the meaning of this term. [3]

d Sketch a diagram of the oxygen molecule showing the outer shell of electrons only. What type of bonding does the molecule contain? [5]

3 Explain the following statements.

a Air is a mixture of elements and compounds. [2]

b The percentage of carbon dioxide in the atmosphere does not significantly vary from 0.04%. [2]

c Power stations are thought to be a major cause of acid rain. [3]

4 Use the words below to complete the following passage about water and its uses.

abundant chlorine monitored coolant reactions pure clay solvent essential

Water is _____ for all life on Earth. It is the most _____ substance on Earth. Water is not only used for drinking and washing but also:

- in chemical _____, for example in the production of ethanol from ethene
- as a _____ to dissolve things
- as a _____ in some chemical processes.

For us to drink it, water must be relatively _____. To make the water drinkable, any undissolved solids are removed through filtration beds. Aluminium sulfate is added to remove small particles of _____ and _____ is added to kill bacteria.

Water is continuously _____ because there are certain substances that eventually find their way into the water supply. [9]

5 a Explain what is meant by the term 'pollution' with reference to air and water. [4]

b i Give an air pollutant produced by the burning of coal. [1]

ii Give a different air pollutant produced by the combustion of petrol in a car engine. [1]

c Some of our drinking water is obtained by purifying river water.

i Would distillation or filtration produce the purest water from river water? Give a reason for your answer. [3]

ii Which process, distillation or filtration, is actually used to produce drinking water from river water? Compare your answer with your answer in **c i**. [2]

6 a Besides carbon dioxide, name one other greenhouse gas. [1]

b Carbon dioxide is a greenhouse gas which adds to the greenhouse effect. Describe what you understand by the term 'greenhouse effect'. [3]

c The graph below shows the change in mean global air temperature from 1860 to 2000.

i Describe the general trend shown by the graph from 1860. [1]

ii Explain what you think has caused the trend you gave as an answer in part **a**. [2]

Organic chemistry 1

You will have seen earlier in Chapter 6 that important substances, called hydrocarbons, are obtained from petroleum. These hydrocarbons are mainly used as fuels of different types. In this chapter you will study two families, or homologous series, of these hydrocarbons, the alkanes and alkenes. You will examine their different structures and different physical and chemical properties. You will see that it is possible to produce alkenes from alkanes in a process called cracking. Finally, you will learn about a very important set of materials that are made from alkenes called polymers, better known as plastics, and the problems that are created by overuse of these polymers.

A lot of the compounds that are present in living things have been found to be compounds containing carbon (Figure 12.1). These are known as **organic compounds**. All living things are made from organic compounds based on chains of carbon atoms which are not only covalently bonded to each other but also covalently bonded to hydrogen, oxygen and/or other elements. The organic compounds are many and varied. Some scientists suggest that there are more than ten million known organic compounds.

▲ **Figure 12.1** Living things contain organic compounds

12.1 Alkanes

Most of the hydrocarbons in petroleum belong to the family of compounds called **alkanes**. The molecules within the alkane family contain carbon atoms covalently bonded to four other atoms by single bonds as shown in Figure 12.2, which shows the fully **displayed formulae** of some alkanes. Because these molecules possess only carbon–carbon single covalent bonds, they are said to be **saturated hydrocarbons**, as no further atoms can be added. This can be seen in the bonding scheme for methane (Figure 12.3). The physical properties of the first six members of the alkane family are shown in Table 12.1 (p. 190).

> **Key definition**
>
> A **displayed formula** shows how the various atoms are bonded and shows all the bonds in the molecule as individual lines.

You will notice from Figure 12.2 and Table 12.1 that the compounds have a similar structure and similar name endings. They also behave chemically in a similar way. A family with these factors in common is called a **homologous series**.

Methane

Ethane

Propane

Butane

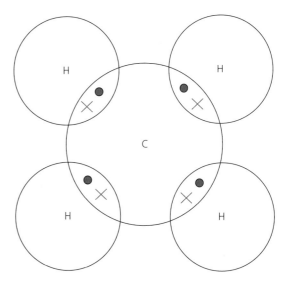

Methane molecule (CH_4)

▲ **Figure 12.3** The covalent bonding scheme for methane

All the members of a homologous series can also be represented by a general formula. In the case of the alkanes, the general formula is:

$$C_nH_{2n+2}$$

where n is the number of carbon atoms present.

As you go up a homologous series, in order of increasing number of carbon atoms, the physical properties of the compounds gradually change. For example, the melting and boiling points of the alkanes shown in Table 12.1 gradually increase. This is due to an increase in the intermolecular forces as the size and mass of the molecule increases (Chapter 3, p. 43). As you can see from Figure 12.2 and Table 12.1, the increase in size and mass of the molecule is due to the addition of a CH_2 group as you descend the homologous series.

The compounds within a homologous series possess similar chemical properties through their **functional group**. A functional group is usually an atom or group of atoms that are present within the molecules of the homologous series. In the case of the alkanes, they are unusual in that they do not have a functional group. For a further discussion of functional groups, see Chapter 13, p. 206.

Under normal conditions, molecules with up to four carbon atoms are gases, those with between five and 16 carbon atoms are liquids, while those with more than 16 carbon atoms are solids.

Pentane

Hexane

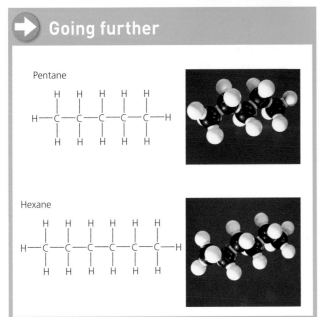

▲ **Figure 12.2** The fully displayed formulae and molecular models of the first six alkanes

Key definitions

A **homologous series** is a family of similar compounds with similar chemical properties due to the presence of the same functional group.

A **functional group** is an atom or group of atoms that determine the chemical properties of a homologous series.

▼ **Table 12.1** Some alkanes and their physical properties

Alkane	Molecular formula	Melting point/°C	Boiling point/°C	Physical state at room temperature
Methane	CH_4	−182	−162	Gas
Ethane	C_2H_6	−183	−89	Gas
Propane	C_3H_8	−188	−42	Gas
Butane	C_4H_{10}	−138	0	Gas

➡ Going further

Alkane	Molecular formula	Melting point/ °C	Boiling point/ °C	Physical state at room temperature
Pentane	C_5H_{12}	−130	36	Liquid
Hexane	C_6H_{14}	−95	69	Liquid

▶ Test yourself

1 Estimate the boiling points for the alkanes with formulae:
  a C_7H_{16}
  b C_8H_{18}.
2 Name the alkane that has the formula C_7H_{16}.

Naming the alkanes

All the alkanes have names ending in -ane. The rest of the name tells you the number of carbon atoms present in the molecule. For example, the compound whose name begins with:

» *meth-* has one carbon atom
» *eth-* has two carbon atoms
» *prop-* has three carbon atoms
» *but-* has four carbon atoms

and so on.

Structural isomerism

Sometimes it is possible to write more than one displayed formula to represent a molecular formula. The displayed formula of a compound shows how the atoms are joined together by the covalent bonds. For example, there are two different compounds with the molecular formula C_4H_{10}. The displayed and structural formulae of these two substances, along with their names and physical properties, are shown in Figure 12.4.

Melting point −138°C

Boiling point 0°C

$CH_3CH_2CH_2CH_3$

a Butane

Melting point −159°C

Boiling point −12°C

$CH_3CH(CH_3)CH_3$

b 2-methylpropane

▲ **Figure 12.4** Displayed and structural formulae for the two isomers of C_4H_{10}

Compounds such as those in Figure 12.4 are known as **structural isomers**. **Isomers** are substances which have the same molecular formula but different **structural formulae** and displayed formulae. The isomers in Figure 12.4 have the same molecular formula, C_4H_{10}, but different structural and displayed formulae. Butane (Figure 12.4a) has the structural formula $CH_3CH_2CH_2CH_3$ and 2-methylpropane (Figure 12.4b) has the structural formula $CH_3CH(CH_3)CH_3$. The brackets show the formula of the branch. The different structures of the compounds shown in Figure 12.4 have different melting and boiling points. Molecule **b** contains a branched chain and has a lower melting point than molecule **a**, which has no branched chain. All the alkane molecules with four or more carbon atoms possess isomers. Perhaps now you can see why there are so many different organic compounds!

> **Key definitions**
>
> **Structural isomers** are compounds with the same molecular formula, but different structural formulae.
>
> The **structural formula** of an organic compound is an unambiguous description of the way the atoms are arranged, including the functional group.

> ### Test yourself
>
> 3 Draw the displayed formulae for the isomers of C_5H_{12}.

For further insight into isomerism, see p. 195 and Chapter 13, p. 207.

12.2 The chemical behaviour of alkanes

Alkanes are rather unreactive compounds. For example, they are generally not affected by alkalis, acids or many other substances. Their most important property is that they burn or combust easily.

Gaseous alkanes, such as methane, will burn in a good supply of air, forming carbon dioxide and water as well as plenty of heat energy.

$$\text{methane} + \text{oxygen} \rightarrow \text{carbon dioxide} + \text{water} + (\Delta H)$$

$$CH_4(g) + 2O_2(g) \rightarrow CO_2(g) + 2H_2O(g)$$

The gaseous alkanes are some of the most useful fuels. As you saw in Chapter 6, they are obtained by the fractional distillation of petroleum. Methane, better known as natural gas, is used for cooking as well as for heating offices, schools and homes (Figure 12.5a). Propane and butane burn with very hot flames and they are sold as liquefied petroleum gas (LPG). In rural areas where there is no supply of natural gas, central heating systems can be run on propane gas (Figure 12.5b). Butane, sometimes mixed with propane, is used in portable blowlamps and in gas lighters.

a This is burning methane

b Central heating systems can be run on propane

▲ **Figure 12.5**

> As stated, alkanes are generally unreactive but, like all homologous series, they will undergo similar chemical reactions. They will react with halogens such as chlorine. Chlorine is quite a reactive non-metal and will react with, for

example, methane in the presence of sunlight or ultraviolet light. This is known as a photochemical reaction, with the UV light providing the activation energy (E_a) for the reaction (see Chapter 6, p. 93). The overall chemical equation for this process is:

methane + chlorine → chloromethane + hydrogen chloride

$$CH_4(g) + Cl_2(g) \rightarrow CH_3Cl(g) + HCl(g)$$

We can see that one hydrogen atom of the methane molecule is substituted by a chlorine

▲ **Figure 12.6** Chloromethane

atom to form chloromethane (see Figure 12.6). This type of reaction is known as a **substitution reaction**. In a substitution reaction, one atom or group of atoms is replaced by another atom or group of atoms: this is know as monosubstitution.

Going further

Chloromethane is used extensively in the chemical industry. For example, chloromethane is used to make silicones. Silicones are used in building and construction. They are able to bond and seal materials such as concrete, glass, granite, plastics and steel (see Figure 12.7). This enables them to work better and last longer. There are dangers associated with substances containing chloromethane. It is a harmful substance and should be treated with caution.

▲ **Figure 12.7** Silicones are used carefully as adhesives as well as sealants

Early anaesthetics relied upon trichloromethane, $CHCl_3$, or chloroform. Unfortunately, this anaesthetic had a

severe problem since the lethal dose was only slightly higher than that required to anaesthetise the patient. In 1956, halothane was discovered by chemists working at ICI. This is a compound containing chlorine, bromine and fluorine which has been used as an anaesthetic in recent years. Its formula is $CF_3CHBrCl$. However, even this is not the perfect anaesthetic since evidence suggests that prolonged exposure to this substance may cause liver damage. The search continues for even better anaesthetics.

A group of compounds called the chlorofluorocarbons (CFCs) were discovered in the 1930s. Because of their inertness they had many uses, especially as a propellant in aerosol cans. CFC-12 or dichlorodifluoromethane, CF_2Cl_2, was one of the most popular CFCs in use in aerosols. Scientists believe that CFCs released from aerosols are destroying the ozone layer and steps are being taken to reduce this threat.

The ozone hole problem

Our atmosphere protects us from harmful ultraviolet radiation from the Sun. This damaging radiation is absorbed by the relatively thin ozone layer found in the stratosphere (Figure 12.8).

Large holes have been discovered in the last 30 years in the ozone layer over Antarctica, Australasia and Europe (Figure 12.8a). Scientists think that these holes have been produced by CFCs such as CFC-12. CFCs escape into the atmosphere and, because of their inertness, remain without further reaction until they reach the stratosphere and the ozone layer. In the stratosphere, the high-energy ultraviolet radiation causes a chlorine atom to split off from the CFC molecule. This chlorine atom, or free radical, then reacts with the ozone.

$$Cl(g) + O_3(g) \rightarrow OCl(g) + O_2(g)$$

This is not the only problem with CFCs. They are also significant greenhouse gases (Chapter 11, p. 183). The ozone depletion and greenhouse effects have become such serious problems that an international agreement known as the Montreal Protocol on Substances that Deplete the Ozone Layer was agreed in 1987.

Research into replacements has taken place since then, producing better alternatives called hydrochlorofluorocarbons (HCFCs). These have lower ozone-depletion effects and are not effective greenhouse gases. The protocols of the 1980s have been superseded by the Kyoto Protocol of 1997 and the 2016 Paris Agreement, which has been signed by 197 countries. It is believed that if the intended agreements are adhered to then the ozone layer will recover by 2050.

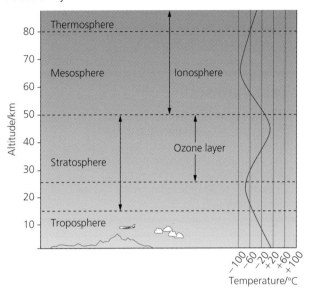

b The ozone layer is between 25 km and 50 km above sea level

a This diagram comes from NASA's ozone monitoring programme, TOMS (Total Ozone Mapping Spectrometer). The ozone hole over the Antarctic (shown in purple and pink on the diagram) is largest in the Antarctic spring. Note: Dobson units are a measure of the total amount of ozone in a vertical column from the ground to the top of the atmosphere

▲ **Figure 12.8**

Other uses of alkanes

Besides their major use as fuels (p. 191), some of the heavier alkanes are used as waxes in candles, as lubricating oils and in the manufacture of another family of hydrocarbons – the alkenes.

> ### ▶ Test yourself
>
> 4 Write a balanced chemical equation to represent the combustion of propane.
> 5 In what mole proportions should chlorine and methane be mixed to produce chloromethane?
> 6 Draw the covalent bonding diagram for chloromethane, CH_3Cl.

Methane – another greenhouse gas

Methane, the first member of the alkanes, occurs naturally. Cows produce it in huge quantities when digesting their food. It is also formed by growing rice. Like carbon dioxide, it is a greenhouse gas (Chapter 11, p. 183) because it acts like the glass in a greenhouse – it will let in heat from the Sun but will not let all of the heat back out again. It is thought that the greenhouse effect may contribute to global warming, which could have disastrous effects for life on this planet. A great debate is going on at the moment as to how we can reduce the amount of methane released into the atmosphere and hence reduce the effects of global warming.

> ### ▶ Test yourself
>
> 7 Use your research skills and your textbook to find out:
>   a any other sources of methane found in nature
>   b how climate change might affect your particular environment.

12.3 Alkenes

Alkenes form another homologous series of hydrocarbons of the general formula C_nH_{2n} where

n is the number of carbon atoms. The alkenes are more reactive than the alkanes because they each contain a double carbon–carbon covalent bond between the carbon atoms (Figure 12.9). Molecules that possess a carbon–carbon double covalent bond, or even a triple covalent bond, are said to be **unsaturated hydrocarbons**, because it is possible to break one of the two bonds to add extra atoms to the molecule. This feature of alkenes is very important and is responsible for the characteristic properties of these organic compounds. This feature is, for the alkenes, known as their functional group.

▲ **Figure 12.9** The bonding in ethene, the simplest alkene

The chemical test to show the difference between saturated and unsaturated hydrocarbons is discussed on p. 197.

Naming the alkenes

All alkenes have names ending in -ene. Alkenes, especially ethene, are very important industrial chemicals. They are used extensively in the plastics industry and in the production of alcohols such as ethanol and propanol. See Table 12.2 and Figure 12.10.

▼ **Table 12.2** The first three alkenes and their physical properties

Alkene	Molecular formula	Melting point/°C	Boiling point/°C	Physical state at room temperature
Ethene	C_2H_4	−169	−104	Gas
Propene	C_3H_6	−185	−47	Gas
Butene	C_4H_8	−184	−6	Gas

Ethene

Propene

Butene

▲ **Figure 12.10** Displayed formula and shape of the first three alkenes

The structural formula of ethene is $CH_2=CH_2$ and the structural formula of propene is $CH_2=CHCH_3$.

Structural isomerism in alkenes

You saw when studying the alkanes that sometimes it is possible to write more than one displayed or structural formula to represent a molecular formula (p. 190). These are known as structural isomers. Butene and the higher alkenes also show structural isomerism. For example, there are three different compounds with the molecular formula C_4H_8. The displayed formulae of two of these substances along with their names and structural formulae are shown in Figure 12.11. The structural formulae of the isomers but-1-ene and but-2-ene give an unambiguous description of the way the atoms are arranged, including the location of the C=C double bond or functional group. The structural formulae are $CH_3CH_2CH=CH_2$ (but-1-ene) and $CH_3CH=CHCH_3$ (but-2-ene).

$CH_3CH_2CH=CH_2$

But-1-ene

$CH_3CH=CHCH_3$

But-2-ene

▲ **Figure 12.11** But-1-ene and but-2-ene: isomers of C_4H_8

Going further

The different isomers shown in Figure 12.11 have different melting and boiling points due to their different structures. All the alkene molecules with four or more carbon atoms possess isomers.

Test yourself

8 Draw the displayed formulae for the isomers of C_5H_{10}.

Where do we get alkenes from?

Very few alkenes are found in nature. Most of the alkenes used by the petrochemical industry are obtained by breaking up larger, less useful alkane molecules obtained from the fractional distillation of petroleum. This is usually done by a process called **catalytic cracking**. In this process the alkane molecules to be 'cracked' (split up) are passed over a mixture of aluminium and chromium oxides heated to 550°C.

$$\text{dodecane} \longrightarrow \text{decane} + \text{ethene}$$

$$C_{12}H_{26}(g) \longrightarrow C_{10}H_{22}(g) + C_2H_4(g)$$

(found in kerosene)   shorter alkane alkene

Another possibility is:

$$C_{12}H_{26}(g) \rightarrow C_8H_{18}(g) + C_4H_8(g)$$

In these reactions, hydrogen may also be formed during cracking. The amount of hydrogen produced depends on the conditions used. Since smaller hydrocarbons are generally in greater demand than the larger ones, cracking is used to match demand (Table 12.3).

▼ **Table 12.3** Percentages of the fractions in petroleum and the demand for them

Fraction	Approx % in petroleum	Approx % demand
Refinery gas	2	5
Gasoline	21	28
Kerosene	13	8
Diesel oil	17	25
Fuel oil and bitumen	47	34

This means that oil companies are not left with large surpluses of fractions containing the larger molecules.

Figure 12.12 shows the simple apparatus that can be used to carry out **thermal cracking** reactions in the laboratory. You will notice that in the laboratory we may use a catalyst of broken, unglazed pottery. This experiment should only ever be carried out as a teacher demonstration.

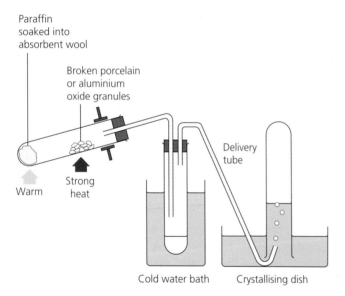

▲ **Figure 12.12** The cracking of an alkane in the laboratory

> **Test yourself**
>
> 9 Using the information in Table 12.2 (p. 194), estimate the boiling point of pentene.
>
> 10 Write a balanced chemical equation to represent the process that takes place when decane is cracked.

12.4 The chemical behaviour of alkenes

The double bond makes alkenes more reactive than alkanes in chemical reactions. For example, hydrogen *adds* across the double bond of ethene, under suitable conditions, forming ethane (Figure 12.13). This type of reaction is called an **addition reaction**. The usefulness of this type of reaction is that only one product is formed.

▲ **Figure 12.13** The addition of hydrogen to ethene using molecular models

Addition reactions

Hydrogenation

Hydrogen adds across the C=C double bond in ethene. This reaction is called hydrogenation. The conditions necessary for this reaction to take place are a temperature of 200°C in the presence of a nickel catalyst.

$$\text{ethene} + \text{hydrogen} \longrightarrow \text{ethane}$$

$$C_2H_4(g) + H_2(g) \longrightarrow C_2H_6(g)$$

Hydrogenation reactions like the one shown with ethene are used in the manufacture of margarines from vegetable oils.

Hydration

Another important addition reaction is the one used in the manufacture of ethanol. Ethanol has important uses as a solvent and a fuel. It is formed when water (as steam) is added across the double bond in ethene in the presence of an acid catalyst. For this reaction to take place, the reactants have to be passed over a catalyst of phosphoric(V) acid (absorbed on silica pellets) at a temperature of 300°C and pressure of 6000 kPa (1 atmosphere = 1×10^5 pascals).

ethene + steam $\underset{\substack{\text{phosphoric(V)} \\ \text{acid catalyst}}}{\overset{300°C, 6000 \text{ kPa}}{\rightleftharpoons}}$ ethanol

$C_2H_4(g)$ + $H_2O(g)$ $\rightleftharpoons$ $C_2H_5OH(g)$

This reaction is reversible as is shown by the equilibrium ($\rightleftharpoons$) sign. The conditions have been chosen to ensure the highest possible yield of ethanol. In other words, the conditions have been chosen so that they favour the forward reaction. The percentage yield is high at approximately 96%. It should be noted, however, that the ethene is a non-renewable resource.

For a further discussion of ethanol and alcohols generally, see Chapter 13, p. 206.

This method of manufacturing ethanol is a continuous process. In a continuous process, reactants are continually fed into the reaction vessel or reactor as the products are removed. Generally, continuous processing is employed if the substance being made is needed on a large scale.

As you have seen, during addition reactions only one product is formed.

Halogenation – a test for unsaturated compounds

The addition reaction between bromine dissolved in an organic solvent, or water, and alkenes is used as a chemical test for the presence of a double bond between two carbon atoms. When a few drops of this bromine solution are shaken with the hydrocarbon, if it is an alkene (such as ethene), a reaction takes place in which bromine joins to the alkene double bond. This results in the bromine solution losing its red/brown colour. If an alkane, such as hexane, is shaken with a bromine solution of this type, no

colour change takes place (Figure 12.14). This is because there are no double bonds between the carbon atoms of alkanes.

ethene + bromine $\longrightarrow$ dibromoethane

$C_2H_4(g)$ + Br_2 (in solution) $\longrightarrow$ $C_2H_4Br_2$ (in solution)

▲ **Figure 12.14** The alkane, on the left, has no effect on the bromine solution, but the alkene, on the right, decolourises it

Test yourself

11 What is meant by the term 'addition reaction'?

12 How would you test the difference between ethane and ethene?

13 Write the displayed formula for pentene.

14 Which of the following organic chemicals are alkanes and which are alkenes?
- Propene, C_3H_6
- Octane, C_8H_{18}
- Nonane, C_9H_{20}
- Butene, C_4H_8

State why you have chosen your answers.

Practical skills

Alkanes and alkenes

For safe experiments/demonstrations which are related to this chapter, please refer to the *Cambridge IGCSE Chemistry Practical Skills Workbook*, which is also part of this series.

Safety

● Eye protection must be worn.

A student wrote the following account of an experiment carried out by the teacher to test which of two liquid hydrocarbons was an alkane and which was an alkene.

To a few drops of each liquid in separate test tubes, they added a few drops of red aqueous bromine solution from a measuring cylinder. They then corked the test tubes and shook them a little. In one test tube the solution went colourless showing this hydrocarbon to be an alkane, while there was no change in colour of the other, so this must be the alkene.

Some statements in the student's account have been underlined by the teacher.

Answer these questions to identify how the experiment could be improved.

1 Which major safety precaution did they forget to mention at the start of the experiment?
2 Do you think the teacher would have added a few drops from a measuring cylinder? Explain your answer.
3 What would be the actual colour of the aqueous bromine solution?
4 How would the teacher have shaken the test tubes?
5 Would you expect only one layer in the solution after adding the aqueous bromine solution? Explain your answer.
6 Has the student got the identification of the alkane and the alkene correct? Explain your answer.

12.5 Polymers

Polythene is a **plastic** that was discovered by accident. Through the careful examination of this substance, when it was accidentally discovered, the plastics industry was born. Polythene is now produced in millions of tonnes worldwide every year. It is made by heating ethene to a relatively high temperature under a high pressure in the presence of a catalyst.

$$n \begin{pmatrix} H \\ \diagdown \\ C = C \\ \diagup \quad \diagdown \\ H \qquad H \end{pmatrix} \longrightarrow \begin{pmatrix} H \quad H \\ | \quad | \\ -C-C- \\ | \quad | \\ H \quad H \end{pmatrix}_n$$

In a chemical equation for this process, we show that a large number of ethene molecules are adding together by using the letter n to indicate very large numbers of monomer molecules. The polymer formed, in the chemical equation, is shown as a repeat unit which repeats n times to form the polymer chain.

When small molecules like ethene join together to form long chains of atoms, called **polymers**, the process is called **polymerisation**. The small molecules, like ethene, which join together in this way are called **monomers**. A polymer chain, a very large molecule or a macromolecule, often consists of many thousands of monomer units and in any piece of plastic there will be many millions of polymer chains. Since in this polymerisation process the monomer units add together to form only one product, the polymer, the process is called **addition polymerisation**.

> **Key definitions**
>
> **Polymers** are large molecules built up from many small units called monomers.
>
> **Plastics** are made from polymers.

▲ **Figure 12.15** This model shows part of the poly(ethene) polymer chain

Poly(ethene), like many other addition polymers has many useful properties including: being tough, easy to mould and an excellent insulator. One of the drawbacks however, as you will see later in this section, is that it is not affected by the weather and does not corrode, and therefore has disposal problems.

It can be found as a substitute for natural materials in plastic bags, sandwich boxes, washing-up bowls, wrapping film, milk-bottle crates and washing-up liquid bottles (Figure 12.16).

▲ **Figure 12.16** These crates are made from poly(ethene)

Other alkene molecules can also produce substances like poly(ethene); for example, propene produces poly(propene), which is used to make ropes and packaging.

In theory, any molecule that contains a carbon-to-carbon double covalent bond can form an addition polymer. For example, ethene, $CH=CH_2$, will undergo an addition reaction to form poly(ethene).

The actual structure of the polymer can only be shown/represented by drawing the monomer as shown in Figure 12.17. In this figure, you will see that the double bond has been replaced by a single bond and there are two extension bonds drawn on either side, which show that the polymer chain extends in both those directions.

Ethene
(monomer)

Repeat unit in
the polymer

▲ **Figure 12.17** Addition polymerisation of ethene

Going further

Other addition polymers

Many other addition polymers have been produced. Often the plastics are produced with particular properties in mind, such as PVC (polyvinyl chloride or poly(chloroethene) and PTFE (poly(tetrafluoroethene)). Both of these plastics have monomer units similar to ethene.

PVC monomer
(vinyl chloride or
chloroethene)

PTFE monomer
(tetrafluoroethene)

If we use chloroethene (Figure 12.18a), the polymer we make is slightly stronger and harder than poly(ethene) and is particularly good for making pipes for plumbing (Figure 12.19).

a Model of chloroethene, the PVC monomer

b Model of part of a PVC polymer chain

▲ **Figure 12.18**

▲ **Figure 12.19** These pipes are made from PVC

PVC is the most versatile plastic and is the second most widely used, after poly(ethene). Worldwide more than 27 million tonnes are produced annually.

If we start from tetrafluoroethene (Figure 12.20a), the polymer we make, PTFE, has some slightly unusual properties:

- it will withstand very high temperatures, of up to 260°C
- it forms a very slippery surface
- it is hydrophobic (water repellent)
- it is highly resistant to chemical attack.

These properties make PTFE an ideal 'non-stick' coating for frying pans and saucepans. Every year more than 50 000 tonnes of PTFE are made.

$$n\left(\begin{array}{c} F \\ | \\ C = C \\ | \\ F \end{array}\begin{array}{c} F \\ | \\ \\ | \\ F \end{array}\right) \longrightarrow \left(\begin{array}{cc} F & F \\ | & | \\ -C - C - \\ | & | \\ F & F \end{array}\right)_n$$

Monomer   Polymer chain

The properties of some addition polymers along with their uses are given in Table 12.4.

a Model of tetrafluoroethene, the PTFE monomer

b Model of part of a PTFE polymer chain

▲ **Figure 12.20**

▼ **Table 12.4** Some addition polymers

Plastic	Monomer	Properties	Uses
Poly(ethene)	$CH_2=CH_2$	Tough, durable	Carrier bags, bowls buckets, packaging
Poly(propene)	$CH_3CH=CH_2$	Tough, durable	Ropes, packaging
PVC	$CH_2=CHCl$	Strong, hard (less flexible than poly(ethene))	Pipes, electrical insulation, guttering
PTFE	$CF_2=CF_2$	Non-stick surface, withstands high temperatures	Non-stick frying pans, soles of irons
Polystyrene	$CH_2=CHC_6H_5$	Light, poor conductor of heat	Insulation, packaging (especially as foam)
Perspex	$CH_2=C(CO_2CH_3)CH_3$	Transparent	Used as a glass substitute

Environmental challenges of plastics

In the last 30 to 40 years, plastics have taken over as replacement materials for metals, glass, paper and wood, as well as for natural fibres such as cotton and wool. This is not surprising since plastics are light, cheap, relatively unreactive, can be easily moulded and can be dyed bright colours. Unfortunately, as you saw earlier in this section, they are not affected by the weather and so do not corrode. This is both an advantage, in the way they can be used, and a disadvantage, as it makes plastics so difficult to dispose of safely. Plastics have contributed significantly to the household waste problem, at least 10% in some countries, and it is getting worse (Figure 12.21)!

▲ **Figure 12.21** This plastic waste is ready to go to landfill

In the recent past, much of our plastic waste has been used as landfill in, for example, disused quarries. However, all over the world these sites are getting very much harder to find and it is becoming more and more expensive to dispose of the waste. Also the older landfill sites are now beginning to produce gases, including methane (which is a greenhouse gas), which contribute to global warming. The alternatives to dumping plastic waste are certainly more economical and more satisfactory but also create their problems.

» Incineration schemes have been developed to use the heat generated from burning waste for heating purposes (Figure 12.22). However, problems with the combustion process (which can result in the production of toxic gases) mean that especially high temperatures have to be employed during the incineration process, driving energy costs up.

» Recycling produces large quantities of black plastic bags and sheeting for resale. However there are problems with recycling in that some plastics cannot be recycled because of their properties.

» **Biodegradable** plastics, as well as those polymers that degrade in sunlight (photodegradable, Figure 12.23a), have been developed. Other common categories of degradable plastics include synthetic biodegradable plastics which are broken down by bacteria, as well as plastics which dissolve in water (Figure 12.23b). The property that allows plastic to dissolve in water has been used in relatively new products, including soluble capsules containing liquid detergent. Very recently, chemists have found ways of involving carbon dioxide and sugars in the production of plastics that are also broken down by bacteria. However, the vast majority of polymers are still non-biodegradable.

There is a further huge global problem which has been highlighted recently. That is, there are extreme pollution problems caused by non-biodegradable plastics and their accumulation in the world's oceans (see Figure 12.24). Aquatic life is being decimated in some regions of the world. The problems with plastics are causing a dramatic rethink of the role of plastics in our society!

▲ **Figure 12.22** A plastic incineration plant

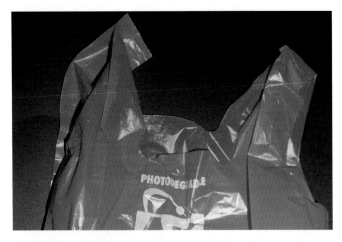

a   This plastic bag is photodegradable

▲ **Figure 12.24** Non-biodegradable plastics accumulate in the oceans of the world

b   This plastic dissolves in water

▲ **Figure 12.23**

> ## Test yourself
>
> **15** Draw the structure of the repeat unit of the addition polymer formed from $CH_3-CH_2CH=CH_2$.
> **16** Draw the displayed formula of the monomer from which the addition polymer below has been produced.
>
> $$\begin{array}{cccccc} C_6H_5 & H & C_6H_5 & H & C_6H_5 & H \\ | & | & | & | & | & | \\ -C & -C & -C & -C & -C & -C- \\ | & | & | & | & | & | \\ H & H & H & H & H & H \end{array}$$

Revision checklist

After studying Chapter 12 you should be able to:
✔ Draw and interpret the displayed formula of a molecule to show all of the atoms and all of the bonds.
✔ Write and interpret general formulae to show the ratio of atoms in the molecules of the compounds in a homologous series.
✔ Identify a functional group as an atom or group of atoms that determine the chemical properties of a homologous series.

✔ State that a structural formula is a description of the way the atoms in a molecule are arranged, including $CH_2=CH_2$.
✔ Define structural isomers as compounds with the same molecular formula, but different structural formulae, including C_4H_{10} as $CH_3CH_2CH_2CH_3$ and $CH_3CH(CH_3)CH_3$, and C_4H_8 as $CH_3CH_2CH=CH_2$ and $CH_3CH=CHCH_3$.
✔ Describe a homologous series as compounds that have the same functional group, the same general formula, differ from one member to the next by a $-CH_2-$ unit , show a trend in physical properties and have similar chemical properties.

✔ State that a saturated compound has molecules in which all carbon–carbon bonds are single bonds.

✔ State that an unsaturated compound has molecules in which one or more carbon–carbon bonds are double bonds or triple bonds.

> ✔ Name and draw the structural and displayed formulae of unbranched alkanes and alkenes and the products of their reactions containing up to four carbon atoms per molecule.

✔ State the type of compound present given the chemical name ending in -ane, -ene, or from a molecular, structural or displayed formula.

✔ State that the bonding in alkanes is single covalent and that alkanes are saturated hydrocarbons.

✔ Describe the properties of alkanes as being generally unreactive, except in terms of combustion and substitution by chlorine.

> ✔ State that in a substitution reaction, one atom or group is replaced by another atom or group.
> ✔ Describe the substitution reaction of alkanes with chlorine as a photochemical reaction, with UV light providing the activation energy, and draw the structural or displayed formulae of the products, limited to monosubstitution.

✔ State that the bonding in alkenes includes a double carbon–carbon covalent bond and that alkenes are unsaturated hydrocarbons.

✔ Describe the manufacture of alkenes and hydrogen by the cracking of larger alkanes using a high temperature, 550°C, and a catalyst.

✔ Describe the reasons for the cracking of larger alkanes.

✔ Distinguish between saturated and unsaturated hydrocarbons using aqueous bromine.

> ✔ State that in an addition reaction, only one product is formed.
> ✔ Describe the properties of alkenes in terms of addition reactions with bromine, hydrogen and steam, and draw the structural or displayed formulae of the products.

✔ Define polymers as large molecules built up from many small units called monomers.

✔ Describe the formation of poly(ethene) as an example of addition polymerisation.

> ✔ Identify the repeat units and/or linkages in polymers.
> ✔ Deduce the structure or repeat unit of an addition polymer from a given alkene and vice versa.

✔ State that plastics are made from polymers.

✔ Describe how the properties of plastics have implications for their disposal.

✔ Describe the environmental challenges caused by plastics in terms of their disposal in landfill, accumulation in oceans and formation of toxic gases from burning.

Exam-style questions

1 Explain the following statements.
   a Ethene is called an unsaturated hydrocarbon. [2]
   b The cracking of larger alkanes into simple alkanes and alkenes is important to the petrochemical industry. [3]

   c The conversion of ethene to ethanol is an example of an addition reaction. [2]

2 The following question is about some of the reactions of ethene.

Gas B

Hydrogen + catalyst A

ETHENE — Burn in air → Gas C + water

Bromine

Steam + catalyst D

Liquid F

Liquid E

   a Give the names and formulae for substances A to F. [12]
   b i Give a word and balanced chemical equation to represent the reaction in which liquid E is formed. [4]
     ii Identify the reaction conditions required for the process to take place. [3]
   c Give the homologous series that gas B belongs to. [1]
   d Describe a chemical test which would allow you to identify gas C. [2]

3 a Petroleum is a mixture of *hydrocarbons* which belong to the *homologous series* called the *alkanes*.
   Explain the meaning of the terms in italics. [6]

   b Alkanes can be converted into substances which are used as solvents. To do this the alkane is reacted with a halogen, such as chlorine, in the presence of ultraviolet light.
      i Give a word and balanced chemical equation for the reaction between methane and chlorine. [4]
      ii Identify the type of reaction taking place. [1]

4 Ethene, C_2H_4, is the starting material for making plastic carrier bags.

$$n\left(\begin{array}{c} H \quad\quad H \\ \diagdown \quad\quad \diagup \\ C = C \\ \diagup \quad\quad \diagdown \\ H \quad\quad H \end{array}\right) \longrightarrow \left(\begin{array}{c} H \quad H \\ | \quad | \\ -C-C- \\ | \quad | \\ H \quad H \end{array}\right)_n$$

   a Identify the type of chemical change taking place in the diagram above. [1]
   b Identify the product formed by this reaction. [1]
   c i The alkene, ethene, is made by cracking large alkane molecules. Describe a simple chemical test to show that ethene is present. [2]

      ii Draw the displayed formula of any products formed in part c i. [1]

5 a The majority of plastic carrier bags are difficult to dispose of.
      i Explain why carrier bags should not just be thrown away. [2]
      ii Explain why plastic carrier bags should be recycled. [2]
      iii Give one advantage that a plastic carrier bag has over one made out of paper. [1]
   b A label like the one below is found on some plastic carrier bags.
      'This plastic carrier bag is made from a substance that is made from the chemical *elements* carbon and hydrogen only. When the carrier bag is burned, it produces carbon dioxide and water. These substances are natural and will not harm the environment.'

i Give the meaning of the term 'element'. [2]
ii Identify the name given to the type of compound that contains the elements carbon and hydrogen only. [1]
iii When the plastic bag burns, heat energy is given out. Give the name used to describe reactions that give out heat energy. [1]
iv The plastic bag will probably give off a toxic gas when it is burned in a limited supply of air. Give the name and formula of this gas. [2]

6 a Identify which of the following formulae represent alkanes, which are alkenes, and which represent neither.

CH_3 $C_{12}H_{24}$
C_6H_{12} $C_{20}H_{42}$
C_5H_{12} C_2H_4
C_6H_6 C_8H_{18}
C_9H_{20} C_3H_7 [3]

b Draw the displayed formulae for all the possible isomers which have the molecular formula C_6H_{14}. [10]

Organic chemistry 2

In Chapter 12, we discussed the organic compounds obtained from petroleum, the alkanes, and how they are converted into another homologous series of hydrocarbons called the alkenes. In this chapter you will learn that if you replace one of the hydrogen atoms on an alkane molecule with a group such as –OH, the hydroxyl group, then you get a new homologous series called the alcohols. Also if you replace one of the hydrogen atoms on an alkane molecule with a –COOH group then you get a homologous series called the carboxylic acids. You will see that whichever group you have attached, it will bring with it a new set of physical and chemical properties. These groups are known as functional groups and it is this group of atoms that is responsible for the characteristic reactions of the organic compound. You will also study the formation and uses of another type of polymer known as condensation polymers. Finally, you will learn about the natural condensation polymers, proteins.

13.1 Functional groups

You saw in Chapter 12 that the functional group of an organic molecule is an atom or group of atoms that determine the chemical properties of a homologous series. It is not surprising, therefore, that there are so many organic molecules given that there are many functional groups that can replace an H atom of basic hydrocarbon molecules.

Table 13.1 shows some examples of functional groups. In the table, R represents an alkyl group or a hydrogen atom. An alkyl group has the general formula based on the alkanes, C_nH_{2n+1}. When $n = 1$, R = CH_3; when $n = 2$, R = C_2H_5, and so on.

▲ **Figure 13.1** This fruit juice contains plenty of vitamin C, or ascorbic acid, which contains the functional group –COOH

▼ **Table 13.1** The functional groups present in some homologous series of organic compounds

Class of compound	Functional group
Alcohols	R–OH
Carboxylic acids	R–COOH
Esters	R–COOR
Amines	R–NH$_2$
Amides	R–CONH$_2$

13.2 Alcohols (R–OH)

The alcohols form another homologous series with the general formula $C_nH_{2n+1}OH$ (or R–OH, where R represents an alkyl group). All the alcohols possess an –OH as the functional group. Table 13.2 shows the names and condensed formulae of the first four members along with their melting and boiling points.

▼ **Table 13.2** Some members of the alcohol family

Alcohol	Formula	Melting point/°C	Boiling point/°C
Methanol	CH_3OH	–94	64
Ethanol	CH_3CH_2OH	–117	78
Propanol	$CH_3CH_2CH_2OH$	–126	97
Butanol	$CH_3CH_2CH_2CH_2OH$	–89	117

Test yourself

1 Which of the following are alcohols?
pentanol, cyclohexane, butene, heptanol

2 What is the name of the alcohol with the chemical formula $C_8H_{17}OH$?

3 Looking at the chemical formulae of the following compounds, which is an alcohol?
C_5H_{12} C_2H_5OH C_4H_{10} C_7H_{14}

The alcohols are named by reference to the corresponding alkane (see Chapter 12, p. 190), always ensuring that the hydrocarbon chain is numbered from the end that gives the lowest number to the position of the –OH group. If the –OH group is positioned at the end of the alcohol, for example in $CH_3CH_2CH_2OH$, then the position of this group is shown by numbering the carbon atom it is attached to as '1'. So this molecule is called propan-1-ol, or propanol for short. If the –OH group is attached to the second carbon atom, $CH_3CH(OH)CH_3$, then this molecule is called propan-2-ol. The displayed formulae of these two isomers of propanol are shown in Figure 13.2 along with those for two isomers of butanol.

Propan-1-ol

Propan-2-ol

a Propan-1-ol and propan-2-ol

Butan-1-ol

Butan-2-ol

b Butanol-1-ol and butan-2-ol

▲ **Figure 13.2** Displayed formulae of alcohols

Alcohols have high boiling points and relatively low volatility. Alcohol molecules are like water molecules (H–OH) in that they are polar (see Chapter 11, p. 173).

Going further

Alcohol molecules are polar because of the presence of the –OH group, in which the hydrogen attached to oxygen creates (within this group) a small difference in charge (Figure 13.3). Other organic molecules that are polar are carboxylic acids such as ethanoic acid (p. 211).

▲ **Figure 13.3** Polar –OH groups in ethanol molecules lead to hydrogen bonding and an attraction between neighbouring molecules

Because of the presence of the polar –OH groups, there is a relatively strong polar attraction between alcohol molecules. This polar attraction between the charges δ+ and δ– in neighbouring molecules is called a hydrogen bond. This means that the molecules have a much stronger attraction for each other than molecules of the corresponding alkane of similar relative molecular mass, M_r. For example, ethanol has an M_r value of 46 and is a liquid at room temperature with a boiling point of 78°C, while propane ($CH_3CH_2CH_3$) has an M_r of 44 and is a gas at room temperature with a boiling point of –42°C.

It is possible to think of alcohol molecules as water molecules in which an H atom has been replaced by an alkyl group, for example, $-C_2H_5$ in C_2H_5OH. This close similarity between their molecules explains why water and alcohols with small molecules, such as methanol and ethanol, are miscible. They mix because of the presence of the polar $-OH$ group in both molecules.

Many other materials, such as food flavourings, are made from ethanol. As ethanol is also a very good solvent and evaporates easily, it is used extensively as a solvent for paints, glues, aftershave and many other everyday products (Figure 13.4).

▲ **Figure 13.4** This aftershave contains alcohol

Ethanol is by far the most important of the alcohols and is often just called 'alcohol'. Ethanol can be produced by fermentation (p. 209) as well as by the hydration of ethene (p. 210). It is a neutral, colourless, volatile liquid which does not conduct electricity.

Combustion

The combustion of ethanol is an important property of ethanol. Ethanol burns quite readily with a clean, hot flame.

$$\text{ethanol} + \text{oxygen} \rightarrow \text{carbon} + \text{water} + (\Delta H) \\ \text{dioxide}$$

$$CH_3CH_2OH(l) + 3O_2(g) \rightarrow 2CO_2(g) + 3H_2O(g) + (\Delta H)$$

As methylated spirit, it is used in in domestic burners for heating and cooking. Methylated spirit is ethanol with small amounts of poisonous substances, such as methanol, added to stop people drinking it. Some countries, like Brazil and the US, already use ethanol

mixed with petrol as a fuel for cars and this use is increasing worldwide. The ethanol used mixed with petrol is also unfit for human consumption.

Oxidation

Vinegar is manufactured by allowing solutions containing alcohol to oxidise. Bacteria present in the solution aid this process to take place. This type of oxidation is how different vinegars are produced. Beer is converted to malt vinegar while wine oxidises to wine vinegar and cider to cider vinegar.

Ethanol can also be oxidised to ethanoic acid (an organic acid also called acetic acid) by powerful oxidising agents, such as acidified potassium manganate(VII). During the reaction, the purple colour of potassium manganate(VII)) is removed (Figure 13.5) as the ethanol is oxidised to ethanoic acid.

$$\text{ethanol} + \text{oxygen} \xrightarrow{\text{heat}} \text{ethanoic} + \text{water} \\ \text{(from} \qquad\qquad \text{acid} \\ \text{potassium} \\ \text{manganate} \\ \text{(VII))}$$

$$CH_3CH_2OH(l) + 2[O] \rightarrow CH_3COOH(aq) + H_2O(l)$$

Carboxylic acid group

▲ **Figure 13.5** Potassium manganate(VII) turns from purple to colourless

Test yourself

4 Why does ethanol make such a good solvent?

Going further

Cholesterol – a complex molecule that contains the –OH group

Cholesterol is a naturally occurring and essential chemical. It belongs to a family of chemicals called steroids and also contains an alcohol group (Figure 13.6). Cholesterol is found in almost all of the tissues in the body, including nerve cells. Levels of cholesterol above normal (above 6.5 mmol/l) are associated with an increased risk of heart disease. Cholesterol hardens and blocks off arteries by building up layers of solid material (atheroma) inside the arteries (Figure 13.7). This is particularly serious if the arteries that supply the heart or brain are blocked. Simple tests are now available to monitor cholesterol levels. People with high levels can be treated and can follow special low-fat and low-cholesterol diets.

a This artery is being blocked by atheroma, which may be related to high levels of cholesterol in the blood

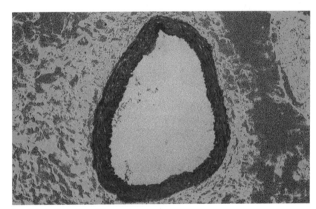

b This is a healthy artery

▲ **Figure 13.7**

▲ **Figure 13.6** The displayed formula of cholesterol

Manufacture of ethanol

Biotechnology involves making use of micro-organisms or their components, such as enzymes, for the benefit of humans to produce, for example, foods such as yoghurt and bread. One of the oldest biotechnologies is that of **fermentation**. It involves a series of biochemical reactions brought about by micro-organisms or enzymes.

Fermentation

Fermentation in the laboratory can be carried out using sugar solution. A micro-organism called yeast is added to the solution. The yeast uses the sugar for energy during **anaerobic respiration** (respiration without oxygen), and so the sugar is broken down to give carbon dioxide and ethanol.

The best temperature for this process to be carried out is 25–35°C. This is because the enzymes in yeast which catalyse this process will be denatured at temperatures much above this, for example, at 40°C.

$$\text{glucose} \xrightarrow{\text{yeast}} \text{ethanol} + \text{carbon dioxide}$$

$$C_6H_{12}O_6(aq) \rightarrow 2C_2H_5OH(l) + 2CO_2(g)$$

The process in industry by which glucose is fermented is a **batch process**. A batch process is one in which each batch goes through one stage of the production process before moving onto the next stage. So in this case, ethanol is the primary product. After fermentation of the glucose has finished, the ethanol in this batch has to be separated from the mixture by fractional distillation. The process then is repeated with another batch of reagents. Batch processes are used where small amounts of a substance are required. The percentage yield by this process is approximately 15%. Figure 13.8 shows a simple apparatus for obtaining ethanol from glucose in the laboratory.

▲ **Figure 13.8** Fermenting glucose and yeast to produce ethanol. The bag is inflated during the experiment by CO_2

Hydration of ethene

Another important addition reaction is the hydration of ethene used in the manufacture of ethanol. Ethanol has important uses as a solvent and a fuel. It is formed when water (as steam) is added across the double bond in ethene in the presence of an acid catalyst. For this reaction to take place, the reactants have to be passed over an acid catalyst (usually phosphoric(V) acid absorbed on silica pellets) at a temperature of 300°C and pressure of 6000 kPa.

$$\text{ethene} \quad + \quad \text{steam} \quad \underset{\substack{\text{phosphoric(V)}\\ \text{acid catalyst}}}{\overset{300°\,C,\ 6000\ kPa}{\rightleftharpoons}} \quad \text{ethanol}$$

$$C_2H_4(g) \quad + \quad H_2O(g) \quad \rightleftharpoons \quad C_2H_5OH(g)$$

For a further discussion of this reaction, see Chapter 12, p. 197.

As discussed earlier, ethanol is a very useful chemical and used in many ways. It can be manufactured by two processes: fermentation of glucose and hydration of ethene. A summary of the methods of manufacture are shown in Table 13.3.

▼ **Table 13.3** Summary of the two methods that are used to make ethanol

	Fermentation	Hydration
Conditions employed	37°C and so less energy needed compared to the hydration method	300°C, 6000 kPa uses a catalyst of phosphoric acid
Processing	Manufactured in batches	Manufactured by a continuous process
Sustainability	Sustainable source – the glucose is renewable	Finite source–ethene is non-renewable
Purification	Fractional distillation	Becomes pure during production
Percentage yield	Low – about 15%	High – about 96%

As you would expect, both methods of preparation of ethanol have their advantages and disadvantages. Table 13.4 shows some of these.

▼ **Table 13.4** Advantages and disadvantages of preparation methods for ethanol

	Advantages	Disadvantages
Fermentation	• Low energy consumption • Uses readily available materials	• Slow process • Not a continuous process
Hydration of ethene	• Much faster process • Continuous process	• High energy consumption • Requires ethene which has to be obtained by cracking

Test yourself

5 Why should being a continuous process be an advantage for a process for making ethanol?

Going further

Baking

To make bread, fresh yeast is mixed with warm sugar solution and the mixture is added to the flour. This dough mixture is then put into a warm place to rise. The dough rises due to the production of carbon dioxide from aerobic respiration (respiration with oxygen) by the yeast. The products of this style of respiration are different to those of anaerobic respiration.

$$\text{sugar} + \text{oxygen} \xrightarrow{\text{yeast}} \begin{array}{c} \text{carbon} \\ \text{dioxide} \end{array} + \text{water} + (\Delta H)$$

$$C_6H_{12}O_6(aq) + 6O_2(g) \rightarrow 6CO_2(g) + 6H_2O(l) + (\Delta H)$$

After the dough has risen, it is baked: the heat kills the yeast and the bread stops rising.

New applications of biotechnology

A large number of firms throughout the world are investing large sums of money in the newer biotechnology applications now in use.

» Enzymes can be isolated from micro-organisms and used to catalyse reactions in other processes. For example, proteases are used in biological detergents to digest protein stains such as blood and food. Also, catalase is used in the rubber industry to help convert latex into foam rubber.

» Our ability to manipulate an organism's genes to make it useful to us is called genetic engineering. This is being used, for example, to develop novel plants for agriculture as well as making important human proteins such as insulin and growth hormone.

However, a word of caution is necessary. The new biotechnologies may not be without dangers. For example, new pathogens (organisms that cause disease) might be created accidentally. Also, new pathogens may be created deliberately for use in warfare. As you can imagine, there are very strict guidelines covering these new biotechnologies, especially in the area of research into genetic engineering.

13.3 Carboxylic acids

The **carboxylic acids** form another homologous series, this time with the general formula $C_nH_{2n+1}COOH$. All the carboxylic acids possess $-COOH$ as their functional group.

Table 13.5 shows the first four members of this homologous series along with their melting and boiling points. Figure 13.9 shows the actual arrangement as displayed formulae of the atoms in these members of this family.

▼ **Table 13.5** Some members of the carboxylic acid series

Carboxylic acid	Structural formula	Melting point/ °C	Boiling point/ °C
Methanoic acid	HCOOH	9	101
Ethanoic acid	CH_3COOH	17	118
Propanoic acid	CH_3CH_2COOH	−21	141
Butanoic acid	$CH_3CH_2CH_2COOH$	−6	164

▲ **Figure 13.9** The displayed formulae of the first four members of the carboxylic acids

Carboxylic acids reacting as acids

Methanoic acid is present in stinging nettles and ant stings. Ethanoic acid, however, is the most well known as it is the main constituent of vinegar. Like other acids, ethanoic acid affects indicators and will react with metals such as magnesium. However, whereas the mineral acids, such as hydrochloric acid, are called strong acids, ethanoic acid is a weak acid (Chapter 8, p. 118). Even though it is a weak acid, it will still react with bases to form salts. For example, the salt sodium ethanoate (CH_3COONa) is formed when ethanoic acid reacts with dilute sodium hydroxide.

ethanoic acid + sodium → sodium + water
                    hydroxide   ethanoate

$$CH_3COOH(aq) + NaOH(aq) \rightarrow CH_3COONa(aq) + H_2O(l)$$

Ethanoic acid also undergoes typical reactions of acids with metals and carbonates. In the case of the metal magnesium, the metal salt magnesium ethanoate (($CH_3COO)_2Mg$) and hydrogen are produced.

ethanoic acid + magnesium → magnesium + hydrogen
                                   ethanoate

$$2CH_3COOH(aq) + Mg(s) \rightarrow (CH_3COO)_2Mg(aq) + H_2(g)$$

Ethanoic acid reacts with carbonates such as sodium, producing the salt sodium ethanoate (CH_3COONa), carbon dioxide and water.

ethanoic acid + sodium → sodium + carbon + water
                 carbonate   ethanoate   dioxide

$$2CH_3COOH(aq) + Na_2CO_3(s) \rightarrow 2CH_3COONa(aq) + CO_2 + H_2O(g)$$

Ethanoic acid also undergoes further typical reactions of acids with indicators turning, for example, methyl orange to red.

Going further

Some interesting carboxylic acids

We come across carboxylic acids in our everyday life. Aspirin is a frequently used painkiller. It is also able to reduce inflammation and fever, and a low dose taken on a daily basis by people over the age of 50 may prevent heart attacks.

Vitamin C, also known as ascorbic acid, is an essential vitamin. Vitamin C is required by the body in very small amounts and it is obtained from foods. It is found in citrus fruits and brightly coloured vegetables, such as peppers and broccoli. Vitamin C prevents the disease scurvy.

Test yourself

6 Describe the reaction that takes place between ethanoic acid and:
   a calcium
   b potassium carbonate
   c potassium hydroxide.
   In each case, write a word equation as well as a balanced chemical equation.

13.4 Esters

Ethanoic acid will react with ethanol, in the presence of a few drops of concentrated sulfuric acid acting as a catalyst, to produce ethyl ethanoate – an **ester.**

ethanoic + ethanol $\underset{H_2SO_4}{\overset{conc}{\rightleftharpoons}}$ ethyl ethanoate + water
acid

$$CH_3COOH(l) + C_2H_5OH(l) \rightleftharpoons CH_3COOC_2H_5(aq) + H_2O(l)$$

This reaction is called esterification.

Members of the 'ester' family have strong and pleasant smells. They have the general formula $C_nH_{2n+1}COOC_xH_{2x+1}$. Esters are named after the acid and alcohol from which they are derived:

» name – alcohol part first, acid part second, e.g. propyl ethanoate
» formula – acid part first, alcohol part second, e.g. $CH_3COOC_3H_7$

Table 13.6 shows the actual arrangement of the atoms in some of the members of this family.

▼ **Table 13.6** The displayed formulae of some esters

Ester	Made from		Structure	3D model
	Alcohol	Carboxylic acid		
Ethyl ethanoate $CH_3COOC_2H_5$	Ethanol C_2H_5OH	Ethanoic acid CH_3COOH		
Propyl methanoate $HCOOC_3H_7$	Propan-1-ol C_3H_7OH	Methanoic acid $HCOOH$		
Methyl butanoate $C_3H_7COOCH_3$	Methanol CH_3OH	Butanoic acid C_3H_7COOH		

Many esters occur naturally and are responsible for the flavours in fruits and the smells of flowers. They are used, therefore, in some food flavourings and in perfumes (Figure 13.10).

Fats and oils are naturally occurring esters which are used as energy storage compounds by plants and animals. They possess the same linkage as PET but have different units (see p. 215).

▲ **Figure 13.10** Perfumes contain esters

Test yourself

7 Which of the following organic chemicals are carboxylic acids or alcohols?
  - Hexanoic acid, $C_5H_{11}COOH$
  - Butanol, C_4H_9OH
  - Octane, C_8H_{18}
  - Nonane, C_9H_{20}
  - Methanoic acid, $HCOOH$
  - Pentene, C_5H_{10}
  - Hexanol, $C_6H_{13}OH$

  State why you have chosen your answers.

8 Write word and balanced chemical equations for the esterification of propanoic acid with ethanol.

Practical skills

For safe experiments/demonstrations which are related to this chapter, please refer to the *Cambridge IGCSE Chemistry Practical Skills Workbook*, which is also part of this series.

Safety

- Eye protection must be worn.

A student wrote the following account of two experiments they observed their teacher carrying out about the properties of carboxylic acids.

1 Some <u>ethanoic acid</u> was poured into a test tube. A small piece of magnesium ribbon was added to the acid. <u>Bubbling</u> took place. The test tube was corked and the gas produced was tested using a lit wooden splint. It gave a squeaky 'pop'. The gas <u>must be oxygen</u>.

2 Some <u>ethanoic acid</u> was poured into a test tube. Some sodium carbonate was <u>added</u> to the acid. <u>Bubbling</u> took place. The test tube was corked and after about 10 seconds the cork was removed and the gas was poured into a test tube containing limewater. This test tube was corked <u>and shaken</u>. The limewater went slightly milky. The gas <u>must be hydrogen</u>.

Some statements in the student's account have been underlined by the teacher.

Answer these questions to identify how the account of the experiments could be improved.

1 In both cases, what strength of ethanoic acid, dilute or concentrated, and volume of acid, 5 cm^3 or 50 cm^3, would be used?
2 In both experiments, what is the more scientifically correct word to use rather than 'bubbling'?
3 What was the gas that gave a squeaky pop?
4 In experiment 2, how is the sodium carbonate added?
5 In experiment 2, how should the shaking take place? Explain your answer.
6 In experiment 2, what was the gas that gave a milky colour with limewater?
7 Are there any other suggestions that you could make to improve the account?

13.5 Condensation polymers

In Chapter 12 (p. 198), you studied the different addition polymers produced from alkenes. Not all polymers are formed by addition reactions though. Some are produced as a result of a different type of reaction. In 1935, Wallace Carothers discovered a different sort of plastic when he developed the thermoplastic, nylon. Nylon is made by reacting two different chemicals together, unlike poly(ethene) which is made only from monomer units of ethene. Poly(ethene), formed by addition polymerisation, can be represented by:

  –A–A–A–A–A–A–A–A–A–A–

where A = monomer.

The starting molecules for nylon are more complicated than those for poly(ethene) and are called 1,6-diaminohexane and hexanedioic acid.

Hexanedioic acid   +   1,6-diaminohexane
$HOOC(CH_2)_4COOH$ $H_2N(CH_2)_6NH_2$

$$H_2N(CH_2)_6CONH(CH_2)_4COOH \ + \ H_2O$$

Amide link

The polymer chain is made up from the two starting molecules arranged alternately (Figure 13.11) due to these molecules reacting and therefore linking up. Each time a reaction takes place, the other product of the reaction, a molecule of water, is lost.

▲ **Figure 13.11** A nylon polymer chain is made up from the two molecules arranged alternately just like the two different coloured poppet beads in the photo

This sort of reaction is called **condensation polymerisation**. This differs from addition polymerisation, where there is only one product. Because an amide link is formed during the polymerisation, nylon is known as a **polyamide**. This is the same amide link as found in proteins (p. 216). It is often called the peptide link. This type of polymerisation, in which two kinds of monomer unit react, results in a chain of the type:

    –A–B–A–B–A–B–A–B–A–B–

Generally, polyamides have the structure:

$$-\overset{O}{\underset{||}{C}}-\blacksquare-\overset{O}{\underset{||}{C}}-\underset{H}{\overset{|}{N}}-\square-\underset{H}{\overset{|}{N}}-\overset{O}{\underset{||}{C}}-\blacksquare-\overset{O}{\underset{||}{C}}-\underset{H}{\overset{|}{N}}-\square-\underset{H}{\overset{|}{N}}-\overset{O}{\underset{||}{C}}-\blacksquare$$

Looking at this structure, it is easy to see that the repeat unit of this polyamide is:

$$-\overset{O}{\underset{||}{C}}-\blacksquare-\overset{O}{\underset{||}{C}}-\underset{H}{\overset{|}{N}}-\square-\underset{H}{\overset{|}{N}}-$$

When nylon is made in industry, it forms as a solid which is melted and forced through small holes (Figure 13.12). The long filaments cool and solid nylon fibres are produced which are stretched to align the polymer molecules, and then dried. The resulting yarn can be woven into fabric to make shirts, ties, sheets and parachutes or turned into ropes or racket strings for tennis and badminton rackets.

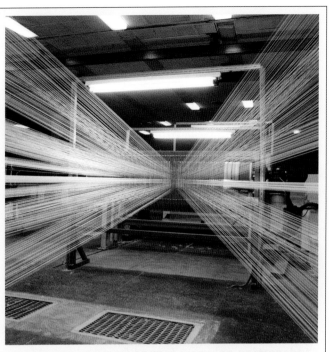

▲ **Figure 13.12** Nylon fibre is formed by forcing molten plastic through hundreds of tiny holes

We can obtain different polymers with different properties if we carry out condensation polymerisation reactions between other monomer molecules. For example, if we react ethane-1,2-diol with benzene-1,4-dicarboxylic acid (terephthalic acid), then we produce a polymer called poly(ethyleneterephthalate) or PET for short.

Ethane-1,2-diol $\qquad$ Benzene-1,4-dicarboxylic acid
$HO(CH_2)_2OH \qquad + \qquad HOOC(C_6H_4)COOH$

$$\downarrow$$

$$HO(CH_2)_2\mathbf{OCO}(C_6H_4)COOH \ + \ H_2O$$

Ester link

This ester link is the same linkage as in fats. Generally, polyesters have the structure:

$$-\overset{O}{\underset{||}{C}}-\blacksquare-\overset{O}{\underset{||}{C}}-O-\square-O-\overset{O}{\underset{||}{C}}-\blacksquare-\overset{O}{\underset{||}{C}}-O-\square-O-$$

Like nylon, PET can be turned into yarn, which can then be woven. PET clothing is generally softer than that made from nylon but both are hard wearing. Because an ester link is formed during the polymerisation, PET is known as a **polyester.**

When this polymer is being used to make bottles, for example, it is usually called PET. An interesting feature of PET is that it can be converted back into its monomers and so can be recycled.

Test yourself

9 Draw the displayed structures of the two monomer units for the following polymer chain.

10 Explain the differences between an addition polymer and a condensation polymer.

13.6 Natural polyamides

Proteins are natural polyamides. They are condensation polymers formed from amino acid monomers. Amino acid monomers have the general structure shown below. R represents different types of side chain.

There are 20 different amino acids and they each possess two functional groups. One is the carboxylic acid group, –COOH. The other is the amine group, –NH$_2$. Two amino acids are glycine and alanine.

Amino acids are the building blocks of proteins. Similar to nylon (see p. 215), proteins are polyamides as they contain the –CONH– group, which is called the amide or, in the case of proteins, the peptide link. Proteins are formed by condensation polymerisation.

A dipeptide
(composed of two amino acids joined together)

Protein chains formed by the reaction of many amino acid molecules have the general structure shown below.

Further reaction with many more amino acids takes place at each end of each molecule to produce the final protein (Figures 13.13 and 13.14). For a molecule to be a protein, there must be at least 100 amino acids involved. Proteins make up some 15% of our body weight and, along with fats and carbohydrates (such as starch and glucose), they are among the main constituents of food.

▲ **Figure 13.13** General structure of a protein

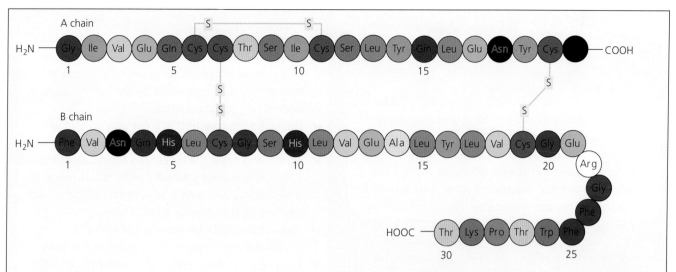

▲ **Figure 13.14** The structure of a protein – human insulin (the different coloured circles represent different amino acids in this protein)

Test yourself

11 How many amino acids have to be involved before the biopolymer is called a protein?

12 a Give the molecular formula of the amino acid valine (see diagram).

  b During a condensation reaction involving this amino acid which two functional groups react?

Revision checklist

After studying Chapter 13 you should be able to:
- ✔ Write the general formulae for the homologous series alcohols and carboxylic acids.

- ✔ Name and draw the structural and displayed formulae of unbranched alcohols, including propan-1-ol, propan-2-ol, butan-1-ol and butan-2ol, and carboxylic acids, as well as the products of their reactions containing up to four carbon atoms per molecule.

- ✔ State the type of compound present given the chemical name ending in -ol or -oic acid or from a molecular, structural or displayed formula.
- ✔ Describe the manufacture of ethanol by fermentation of glucose and the catalytic addition of steam to ethene.
- ✔ Describe the advantages and disadvantages of the two methods for manufacturing ethanol.
- ✔ Describe the combustion of alcohols.
- ✔ State the uses of ethanol as a solvent and a fuel.
- ✔ Describe the reactions of carboxylic acids with metals, bases and carbonates including names and formulae of the salts produced.

- ✔ Describe the formation of ethanoic acid by the oxidation of ethanol with acidified potassium manganate(VII) as well as by bacterial oxidation during vinegar production.
- ✔ Describe the reaction of a carboxylic acid with an alcohol to give an ester.
- ✔ Name and draw the displayed formulae of the unbranched esters which can be made from unbranched alcohols and carboxylic acids, each containing up to four carbon atoms.
- ✔ Deduce the structure or repeat unit of a condensation polymer from given monomers.
- ✔ Describe the differences between condensation and addition polymerisation.
- ✔ Describe and draw the structure of nylon (a polyamide) and PET (a polyester).
- ✔ Describe proteins as natural polyamides.
- ✔ Draw the general structure of a natural polyamide.
- ✔ Describe and draw the structure of proteins.

Exam-style questions

1 a The *functional group* of the *homologous series* alcohols is –OH. Explain what is meant by the terms in italics. [4]

b Give the general formula of the homologous series of alcohols. [1]

c Butan-1-ol is used as a solvent for paints. It can be made by the hydration of but-1-ene.
  i Give an equation for the reaction that takes place to produce butan-1-ol. [3]
  ii Identify the reaction conditions for this reaction to take place. [3]

d Butan-1-ol will undergo bacterial oxidation if left in the air. Give the name of the organic substance formed by this oxidation process and give its formula. [2]

2 A piece of cheese contains protein. Proteins are natural polymers made up of amino acids. There are 20 naturally occurring amino acids. The displayed formulae of two amino acids are shown below.

Glycine   Alanine

$H_2N-\overset{\overset{H}{|}}{\underset{\underset{H}{|}}{C}}-COOH$ $H_2N-\overset{\overset{H}{|}}{\underset{\underset{CH_3}{|}}{C}}-COOH$

a Identify the type of polymerisation involved in protein formation. [1]

b Give the molecular formulae of the two amino acids shown. [2]

c Sketch a displayed formula to represent the product formed by the reaction between the amino acids shown above. [2]

3 a Copy the following table and complete it by giving the displayed formulae for ethanol and ethanoic acid.

Ethane	Ethanol	Ethanoic acid
$H-\overset{\overset{H}{|}}{\underset{\underset{H}{|}}{C}}-\overset{\overset{H}{|}}{\underset{\underset{H}{|}}{C}}-H$		

[4]

b Describe a simple chemical test that could be used to distinguish ethanol from ethanoic acid. [3]

c i Identify the class of compound produced when ethanol reacts with ethanoic acid. [1]
  ii What catalyst is required for this reaction? [1]
  iii Give a word and balanced chemical equation for this reaction. [4]

4 a Ethanoic acid is a weak acid. Describe what you understand by the term weak acid. [1]

b Ethanoic acid will react with sodium carbonate. Write a word and balanced chemical equation for the reaction. [4]

c Ethanoic acid will react with metals such as magnesium. Write a word and balanced chemical equation for the reaction. [4]

d Ethanoic acid will react with bases such as sodium hydroxide.
  i What do you understand by the term 'base'? [2]
  ii Give another example of a base. [1]
  iii Write a word and balanced chemical equation for the reaction. [4]

5 a Identify which carboxylic acids and alcohols you use to make the following esters:
  i butyl methanoate [2]
  ii ethyl methanoate [2]
  iii propyl propanoate. [2]

b Identify which esters would be made using the following carboxylic acids and alcohols:
  i ethanoic acid and butan-1-ol [1]
  ii methanoic acid and methanol [1]
  iii propanoic acid and butan-1-ol. [1]

c Sketch the displayed formulae of the following esters:
  i ethyl ethanoate [2]
  ii propyl ethanoate. [2]

6 a Identify the polymerisation process that is used to make both nylon and PET. [1]

b Give the starting materials for making:
  i nylon [2]
  ii PET. [2]

c Give the name and formula of the small molecule produced during the polymerisation reactions to produce both nylon and PET. [2]

d Give the name of the chemical link that holds together:
  i nylon [1]
  ii PET. [1]

Experimental techniques and chemical analysis

This chapter shows how to safely and accurately use different experimental techniques, apparatus and materials. This will aid your ability to plan experiments and investigations.

You will also learn that there are specific criteria for purity of substances. Finally, you will learn about the branch of chemistry that deals with the identification of elements, ions or grouping of elements as ions present in a sample, which is called qualitative chemical analysis, or qualitative analysis for short. The various tests you will study would allow you to carry out a wide range of qualitative analysis.

14.1 Apparatus used for measurement in chemistry

Scientists find out about the nature of materials by carrying out experiments in a laboratory. Many of these experiments require apparatus, such as that you have used in your study of chemistry to date. A knowledge and understanding of the use of this scientific apparatus is required for successful experimentation and investigations. In an experiment, you will first have to decide on the measurements to be made and then collect together the apparatus and materials required. The quantities you will need to measure most often in a chemistry laboratory will be temperature, mass, volume of liquids as well as gases, and time.

There will be advantages and disadvantages of each method or piece of apparatus, so when you choose which to use, you will have to use your knowledge of the apparatus to decide if it will be suitable. You will have to decide:

» What apparatus should you use to measure each of these? For example, it sounds silly but you would not use a thermometer to measure the timing of a particular reaction rate!
» Which piece of apparatus is most suitable for the task in hand? For example, you would not use a large gas syringe to measure out 10 cm³ of liquid: you would use a small measuring cylinder because

it gives you a more accurate measurement of volume and it is easier to use.
» How do you use the piece of apparatus correctly? Before you use a piece of apparatus, be sure you know how to use it properly. You need to be safe in your working habits and also to ensure you use good techniques, for example, measuring volume of solutions carefully, so that you make and record measurements accurately. If necessary, ask your teacher before you use a particular piece of apparatus if it is unfamiliar to you.

Measurement of time

▲ **Figure 14.1** This stopwatch can be used to measure the time passed in a chemical reaction. The reading is 6.3 s

Experiments involving rates of reaction will require the use of an accurate **stopwatch** – one that measures to a hundredth of a second. The units of time are hours (h), minutes (min) and seconds (s).

Measurement of temperature

▲ **Figure 14.2** A thermometer can be used to measure temperature. The reading is 3°C

The most commonly used thermometers in a laboratory are alcohol-in-glass. Mercury in-glass thermometers can be used but should be handled with great care. The mercury inside them is poisonous and should not be handled if a thermometer breaks. The units of temperature are those of the Celsius scale. This scale is based on the temperature at which water freezes and boils, that is:

» the freezing point of water is 0°C
» the boiling point of water is 100°C.

The usual thermometer used is that shown in Figure 14.2, which measures accurately between –10° and 110°C at 1°C intervals. It may also be possible to read the thermometer to the nearest 0.5°C if the reading is between two of the scale marks. When reading the thermometer, always ensure that your eye is at the same level as the liquid meniscus in the thermometer to ensure there are no parallax effects (Figure 14.3). The meniscus is the way that the liquid curves at the edges of the capillary in which the liquid is held in the thermometer.

Eye level high –
temperature read 18.6°C

Correct eye level –
temperature read 17.5°C

Eye level low –
temperature read 16.5°C

▲ **Figure 14.3** Eye should be level with the liquid meniscus

Measurement of mass

▲ **Figure 14.4** An electronic balance can be used to measure the mass of reagents. The reading is 155.12 g

There are many different electronic balances which can be used. The precision of an electronic balance is the size of the smallest mass that can be measured on the scale setting you are using, usually 0.01 g. The units for measuring mass are grams (g) and kilograms (kg): 1 kg = 1000 g.

When using an electronic balance, you should wait until the reading is steady before taking it.

Measurement of volume of liquids

▲ **Figure 14.5** The apparatus shown in the photograph (left to right: volumetric pipette, measuring cylinder and burette) are generally used in different experiments to measure volume accurately

Different experiments involving liquids will require one or other or all the various measuring apparatus available for measuring volume. Figure 14.5 shows three of the most commonly used measuring apparatus.

Acid-base titration

A burette and a pipette are needed in a titration. This technique is often used to test the strength or purity of an acid or an alkali. A titration involves the slow, careful addition of one solution, such as an acid, of a known concentration from a burette to a known volume of another solution, such as an alkali, of unknown concentration held in a flask. The solution in the burette is known as a **titrant**. This addition of acid continues until the reaction reaches neutralisation, which is indicated by a colour change in an indicator. For a further discussion of indicators, see Chapter 8 (p. 117). The known volume in the flask would require the use of a pipette. For example, in an acid-alkali titration, the acid would be the titrant. For a further discussion of titrations, see Chapter 8, p. 128.

A measuring cylinder (sometimes called a mixing or graduated cylinder) is a common piece of laboratory equipment used to measure the volume of a liquid. It has a narrow cylindrical shape. Each marked line on the graduated cylinder shows the amount of liquid that has been measured.

The volume of a liquid is a measure of the amount of space that it takes up. The units of volume are litres (l), cubic decimetres (dm^3) and cubic centimetres (cm^3).

$$1 \text{ litre} = 1 \text{ dm}^3 = 1000 \text{ cm}^3$$

However, some of the manufacturers of apparatus used for measuring volume use millilitres (ml). This is not a problem, however, since $1 \text{ cm}^3 = 1 \text{ ml}$.

When reading the volume using one of these pieces of apparatus, it is important to ensure that the apparatus is vertical and that your eye is level with the meniscus of the liquid being measured, as shown in Figure 14.6. The precision of the measurement will vary depending on the apparatus used. For example, if the smallest scale division on a burette is 0.1 cm^3, the precision of any volume measured with the burette will be 0.1 cm^3. It may

also be possible to read the burette to the nearest 0.05 cm^3 if the bottom of the meniscus is between divisions.

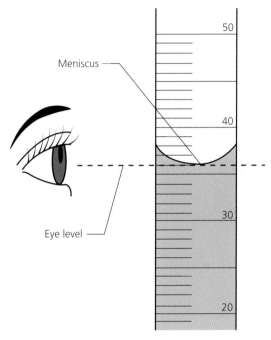

▲ **Figure 14.6** The volume level in this measuring cylinder should be read on the dotted line. The reading here is 36 cm^3

Measurement of volume of gases

The volume of a gas can be measured with a gas syringe. This is used to measure the amount of gas collected in experiments. They have a maximum volume of 100 cm^3.

▲ **Figure 14.7** A gas syringe

For example, in certain reactions (see p. 106) the reaction rate can be followed by collecting the volume of gas generated with time. The gas syringe is connected by a tube to the flask that is giving off the gas as shown in Figure 14.8.

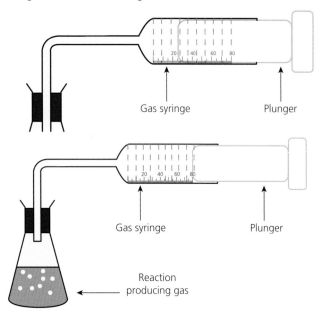

▲ **Figure 14.8** As gas is collected during the reaction, the plunger is forced out and the volume can be read from the scale on the side

The gas syringe has a scale along its length that allows the volume of the gas collected to be measured. As the volume of gas collected increases, the plunger moves further out of the syringe. The volume of collected gas can be measured in a given time, as it increases, and a graph of volume against time can be plotted. This can then be used to show the rate of the reaction (see Chapter 7).

b What are the readings on the two balances?

c What are the readings on these three thermometers?

d Why must mercury thermometers be handled with care?

14.2 Separating mixtures

Many mixtures contain useful substances mixed with unwanted material. To obtain these useful substances, chemists often have to separate them from the impurities. Chemists have developed many different methods of separation, particularly for separating compounds from complex mixtures. Which separation method they use depends on what is in the mixture and the properties of the substances present. It also depends on whether the substances to be separated are solids, liquids or gases.

Test yourself

1 a i Copy and complete the following about the units of volume measurement.
        1 dm^3 = _____ cm^3
      ii What are the readings on the following measuring cylinders?

Separating solid/liquid mixtures

If a solid substance is added to a liquid, it may dissolve to form a **solution**. In this case, the solid is said to be soluble and is called the **solute**. The liquid it has dissolved in is called the **solvent**. An example of this type of process is when sugar is added to tea or coffee. What other examples can you think of where this type of process takes place?

Sometimes the solid does not dissolve in the liquid. This solid is said to be insoluble. For example, tea leaves do not dissolve in boiling water when tea is made from them, although the soluble materials from which tea is made are seen to dissolve from them.

> **Key definitions**
>
> A **solvent** is a substance that dissolves a solute.
>
> A **solute** is a substance that is dissolved in a solvent.
>
> A **solution** is a liquid mixture composed of two or more substances.

Filtration

When you pour a cup of tea through a tea strainer, you are carrying out a filtering process. **Filtration** is a common separation technique used in chemistry laboratories throughout the world. It is used when a solid needs to be separated from a liquid. For example, sand can be separated from a mixture with water by filtering through filter paper as shown in Figure 14.9.

The filter paper contains holes that, although too small to be seen, are large enough to allow the molecules of water through but not the sand particles. It acts like a sieve. The sand gets trapped in the filter paper and the water passes through it. The sand is called the **residue** and the water is called the **filtrate**.

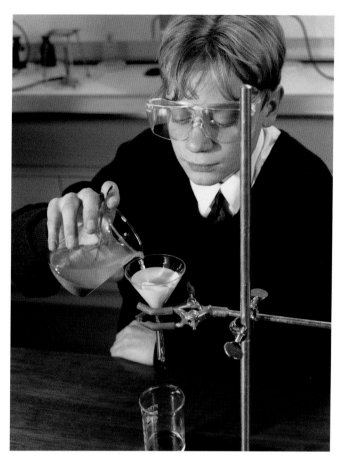

▲ **Figure 14.9** It is important when filtering not to overfill the filter paper

> **Key definitions**
>
> **Residue** is a substance that remains after evaporation, distillation, filtration or any similar process.
>
> **Filtrate** is a liquid or solution that has passed through a filter.

Going further

Decanting

Vegetables do not dissolve in water. When you have boiled some vegetables, it is easy to separate them from the water by pouring it off. This process is called **decanting**. This technique is used quite often to separate an insoluble solid, which has settled at the bottom of a flask, from a liquid.

Centrifuging

Another way to separate a solid from a liquid is to use a centrifuge. This technique is sometimes used instead of filtration. It is usually used when the solid particles are so small that they spread out (disperse) throughout the liquid and remain in suspension. They do not settle to the bottom of the container, as heavier particles would do, under the force of gravity. The technique of centrifuging or centrifugation involves the suspension being spun round very fast in a centrifuge so that the solid gets flung to the bottom of the tube (Figure 14.10).

Spins round

Liquid

Insoluble solids are thrown to the bottom

a The sample is spun round very fast and the solid is flung to the bottom of the tube

b An open centrifuge

▲ **Figure 14.10**

▲ **Figure 14.11** Whole blood (top) is separated by centrifuging into blood cells and plasma (bottom)

The pure liquid can be decanted after the solid has been forced to the bottom of the tube. This method of separation is used extensively to separate blood cells from blood plasma (Figure 14.11). In this case, the solid particles (the blood cells) are flung to the bottom of the tube, allowing the liquid plasma to be decanted.

Evaporation

If a solid has dissolved in a liquid, it cannot be separated by filtering or centrifuging. Instead, the solution can be heated so that the liquid **evaporates** completely and leaves the solid behind. For example, the simplest way to obtain salt from its solution is by slow evaporation, as shown in Figure 14.12.

▲ **Figure 14.12** Apparatus used to slowly evaporate a solvent

You may need to choose a suitable solvent to separate a mixture. For example, if one of the substances in a mixture is soluble in water, you could use water to dissolve the soluble substance, filter the mixture to separate the insoluble substance, and then evaporate the water to retrieve the soluble substance.

Crystallisation

In many parts of the world, salt is obtained from sea water on a vast scale. This is done by using the heat of the Sun to evaporate the water to leave a saturated solution of salt known as brine. A **saturated solution** is defined as one that contains the maximum concentration of a solute dissolved in the solvent. It is important to be aware that solubility of a substance varies with temperature, so the maximum concentration will vary at different temperatures. More of the solute will dissolve at a higher temperature. When the solution is saturated, the salt begins to **crystallise** and it is removed using large scoops (Figure 14.13).

> **Key definition**
> A **saturated solution** is a solution containing the maximum concentration of a solute dissolved in the solvent at a specified temperature.

▲ **Figure 14.13** Salt is obtained in north-eastern Brazil by evaporation of sea water

Simple distillation

If we want to obtain the solvent from a solution, then the process of **distillation** can be carried out. The apparatus used in this process is shown in Figure 14.14.

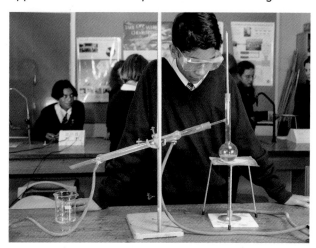

▲ **Figure 14.14** Water can be obtained from salt water by distillation

Water can be obtained from salt water using this method. The solution is heated in the flask until it boils. The steam rises into the Liebig condenser, where it condenses back into water. The salt is left behind in the flask.

Going further

In hot and arid countries, such as Saudi Arabia, this sort of technique is used on a much larger scale to obtain pure water for drinking (Figure 14.15). This process is carried out in a desalination plant.

▲ **Figure 14.15** This desalination plant produces large quantities of drinking water in Saudi Arabia

Separating liquid/liquid mixtures

In recent years, there have been several oil tanker accidents that have led to environmental disasters, just like the one shown in Figure 14.16. These have resulted in millions of litres of oil being washed into the sea. Oil and water do not mix easily. They are said to be **immiscible**. When cleaning up disasters of this type, a range of chemicals can be added to the oil to make it more soluble. This results in the oil and water mixing with each other. They are now said to be **miscible**. The following techniques can be used to separate mixtures of liquids.

▲ **Figure 14.16** Millions of litres of oil can be spilled from oil tankers and cleaning up is a slow and costly process

Immiscible liquids

If two liquids are immiscible, they can be separated using a separating funnel. The mixture is poured into the funnel and the layers allowed to separate. The lower layer can then be run off by opening the tap as shown in Figure 14.17.

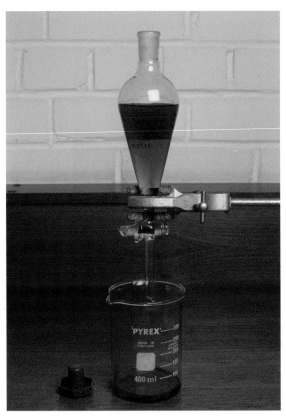

▲ **Figure 14.17** The blue liquid is more dense than the red liquid and so sinks to the bottom of the separating funnel. When the tap is opened, the blue liquid can be run off.

Miscible liquids

If miscible liquids are to be separated, then this can be done by fractional distillation. The apparatus used for this process is shown in Figure 14.18 and could be used to separate a mixture of ethanol and water.

Fractional distillation relies upon the liquids having different boiling points. When an ethanol and water mixture is heated, the vapours of ethanol and water boil off at different temperatures and can be condensed and collected separately.

Ethanol boils at 78°C whereas water boils at 100°C. When the mixture is heated, the vapour produced is mainly ethanol with some steam. Because water has the higher boiling point of the two, it condenses out from the mixture with ethanol. This is what takes place in the fractionating column. The water condenses and drips back into the flask while the ethanol vapour moves up the column and into the condenser, where it condenses into liquid ethanol and is collected in the receiving flask as the **distillate**. When all the ethanol has distilled over, the temperature reading on the thermometer rises steadily to 100°C, showing that the steam is now entering the condenser. At this point the receiver can be changed and the condensing water can now be collected.

Thermometer

Condenser

Cooling water in

Cooling water out

Fractionating column with short lengths of glass rod inside (increases surface area)

Flask

Flask

Liquid mixture

Distillate

Heat

Support

▲ **Figure 14.18** Typical fractional distillation apparatus

Fractional distillation is used to separate miscible liquids such as those in petroleum (crude oil) (see p. 89, Figure 6.4a and b), and the technique can also separate individual gases, such as nitrogen, from the mixture we call air (Figure 14.19).

▲ **Figure 14.19** Gases from the air are extracted in this fractional distillation plant

Going further

Separating solid/solid mixtures

You saw in Chapter 2 (see p. 19, Figure 2.9) that it was possible to separate iron from sulfur using a magnet. In that case, we were using one of the physical properties of iron, that is, the fact that it is magnetic. In a similar way, it is possible to separate scrap iron from other metals by using a large electromagnet like the one shown in Figure 14.20.

▶ **Figure 14.20** Magnetic separation of iron-containing materials

It is essential that, when separating solid/solid mixtures, you pay particular attention to the individual physical properties of the components. If, for example, you wish to separate two solids, one of which sublimes, then this property should dictate the method you employ.

In the case of an iodine/common salt mixture, the iodine sublimes but salt does not. Iodine can be separated by heating the mixture in a fume cupboard as shown in Figure 14.21. The iodine sublimes and re-forms on the cool inverted funnel.

▲ **Figure 14.21** Apparatus used to separate an iodine/salt mixture. The iodine sublimes on heating

Solvent extraction

Sugar can be obtained from crushed sugar cane by adding water. The water dissolves the sugar from the sugar cane (Figure 14.22). This is an example of solvent extraction. In a similar way, some of the green substances can be removed from ground-up grass using ethanol. The substances are extracted from a mixture by using a solvent which dissolves only those substances required.

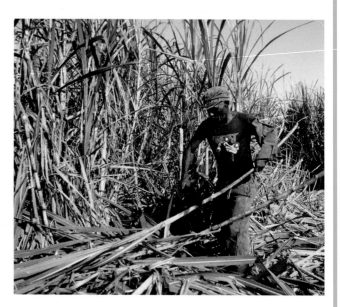

▲ **Figure 14.22** Cutting sugar cane, from which sugar can be extracted by using a suitable solvent

> ### ▶ Test yourself
>
> 2 Which experimental technique(s)/processes would you use to obtain:
>   a the silver bromide precipitate produced from the reaction of silver nitrate solution with sodium bromide solution?
>   b pure water from an aqueous solution of iron(II) sulfate?
>   c common salt from a mixture of the solids salt and sand?
>   d nitrogen gas (boiling point –195.8°C) from liquid air?
>   e calcium chloride solid from calcium chloride solution?

Chromatography

Sometimes we need to separate two or more solids that are soluble; for example, mixtures of soluble coloured substances such as inks and dyes. A technique called **chromatography** is widely used to separate these materials so that they can be identified.

There are several types of chromatography; however, they all follow the same basic principles. The simplest kind is paper chromatography. To separate the different coloured dyes in a sample of black ink, a pencil line is drawn about 1 cm from the bottom of a piece of chromatography paper. This is called the baseline. Do not use a pen to draw the baseline as the dyes in its ink will interfere with the results. A spot of the ink is put onto the pencil line on the chromatography paper. This paper is then put into a suitable container, such as a beaker, containing a suitable solvent, ensuring that the solvent level does not reach the pencil line as shown in Figure 14.23. A lid is placed over the container to reduce evaporation of the solvent.

a Chromatographic separation of black ink

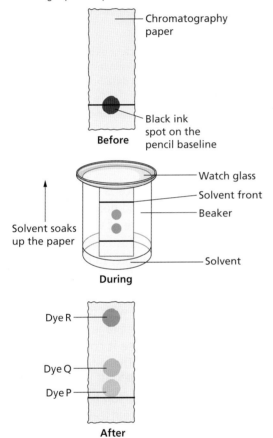

- Chromatography paper
- Black ink spot on the pencil baseline

Before

- Watch glass
- Solvent front
- Beaker
- Solvent

Solvent soaks up the paper

During

Dye R
Dye Q
Dye P

After

b The black ink separates into three dyes: P, Q and R

▲ **Figure 14.23**

As the solvent moves up the paper, the dyes are carried with it and begin to separate. They separate because the substances have different solubilities in the solvent and are absorbed to different degrees by the chromatography paper. As a result, they are separated gradually as the solvent moves up the paper. The chromatogram in Figure 14.23b shows how the ink contains three dyes, P, Q and R.

Chromatography is used extensively in medical research and forensic science laboratories to separate a variety of mixtures.

? Worked example

Several different dyes were spotted on chromatography paper and set up for chromatography. The chromatogram produced is shown below.

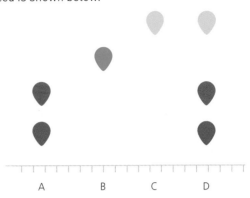

A   B   C   D

Which of the dyes are pure substances and which are mixtures?

The single spots show they are a single dye. So dyes B and C are single dyes.

The dyes that have produced multiple spots are mixtures. So dyes A and D are mixtures of dyes.

→ Going further

Paper chromatography of the products of starch hydrolysis

This technique can be used to identify the products of a reaction. For example, starch can be broken up using enzymes in water solution or water in dilute acid. This type of reaction is called hydrolysis. However, the different methods produce different products, as shown in Figure 14.24.

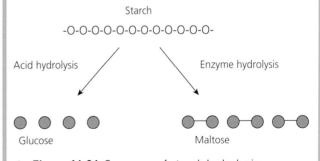

▲ **Figure 14.24** Summary of starch hydrolysis

The products of these reactions can be identified using chromatography. The chromatography paper is 'spotted' with the samples as well as pure glucose and maltose. When this is developed after some hours in the solvent, the spots may be identified (Figure 14.25).

Chromatography paper

Beaker

Origin (start)

Solvent

Hydrolysis A

Glucose

Maltose

Hydrolysis B

After several hours in the solvent, the chromatogram is sprayed with a substance that makes the sugars stand out. It then looks like this.

A

B

Glucose

Maltose

▲ **Figure 14.25** The products of hydrolysis may be identified by chromatography

? Worked example

Which substance has been produced by which hydrolysis in Figure 14.25?

Acid hydrolysis – sample A on the chromatogram. That spot has risen to the same height as the substance glucose.

Enzyme hydrolysis – sample B on the chromatogram. That spot has risen to the same height as the substance maltose.

Numerical measurements (retardation factors) known as R_f **values** can be obtained from chromatograms. An R_f value is defined as the ratio of the distance travelled by the solute (for example, P, Q or R) to the distance travelled by the solvent from the pencil line.

$$R_f = \frac{\text{distance travelled by solute}}{\text{distance travelled by solvent}}$$

The substances to be separated do not have to be coloured. Colourless substances can be made visible by spraying the chromatogram with a **locating agent**. The locating agent will react with the colourless substances to form a coloured product. In other situations, the position of the substances on the chromatogram may be located using ultraviolet light.

? Worked example

The following values were obtained for the chromatogram produced for the amino acid glycine.

Distance travelled by glycine = 2.5 cm

Distance travelled by the solvent = 10.0 cm

$$R_f = \frac{\text{distance travelled by solute}}{\text{distance travelled by solvent}}$$

$$R_f = \frac{2.5}{10.0} = 0.25$$

R_f values are catalogued in data books for a vast array of substances. So R_f values of unknown substances can be compared with values inside a data book to identify the unknown substance.

Criteria for purity

In chemistry, a pure substance is one which contains a single element or compound not mixed with any other substance. The measure of whether a substance is pure is known as purity. This is different to the everyday meaning. For example, in the food industry, if bottles or cartons which are labelled as 'pure' apple juice are sold in shops then the label means that the contents are just apple juice with no other substances being added. However, the apple juice is not pure in the chemical sense because it contains different substances mixed together. The substances added here may include a preservative as well as other substances which are harmless to the person buying it.

However, in the pharmaceuticals industry, purity is essential because there could be significant dangers to health if any other substances were found in medicines. Drugs (pharmaceuticals) are manufactured to a very high degree of purity (Figure 14.26). To ensure that the highest possible purity is obtained, the drugs are dissolved in a suitable solvent and subjected to fractional crystallisation.

Purity can have an important effect on the chemical properties of a substance. Pure substances will form predictable products in chemical reactions. This reliability comes from the fact that there are no impurities present that could start other reactions. Reactions due to impurities may interfere with the reaction that is being studied by a chemist, so chemists often use substances which have a high degree of purity when conducting chemical research.

Throughout the pharmaceutical, food and chemical industries, it is essential that the substances used are pure. The purity of a substance can be assessed by:

» its melting point – if it is a pure solid, it will have a sharp melting point. If an impurity is present then melting takes place over a range of temperatures to show it is a mixture of two or more substances.
» its boiling point – if it is a pure liquid, the temperature will remain steady at its boiling point. If the substance is impure then the mixture will boil over a temperature range.
» chromatography – if it is a pure substance, it will produce only one well-defined spot on a chromatogram. If impurities are present then several spots will be seen, as on the chromatogram in Figure 14.23, p. 231.

▲ **Figure 14.26** These pharmaceuticals must have been through a lot of testing for purity before they can be sold in a pharmacy

Test yourself

3 The following values were obtained for the chromatogram produced for the amino acid proline.

Distance travelled by proline = 6.0 cm

Distance travelled by the solvent = 10.0 cm

Calculate the R_f value for proline.

14.3 Qualitative analysis

The branch of chemistry that deals with the identification of elements or grouping of elements present in a sample is called qualitative chemical analysis, or qualitative analysis for short. It does not deal with quantities.

The techniques employed in qualitative analysis vary in their complexity, depending on the nature of the sample under investigation. In some cases, it is only necessary to confirm the presence of certain elements or groups for which specific chemical tests, or 'spot' tests, applicable directly to the sample, may be available. More often, the sample is a complex mixture, and a systematic analysis must be made in order that all the component parts may be identified. Often, the first simple stages of qualitative analysis require no apparatus at all. Things like colour and smell can be observed without any need for apparatus.

The following summary collects together information from throughout the book which would allow you to carry out qualitative analysis. The material which is of extreme importance as far as your revision is concerned is that given in the following tables.

» Table 14.2 Characteristic flame colours of some metal ions
» Table 14.3 Effect of adding sodium hydroxide solution to solutions containing various metal ions
» Table 14.4 Effect of adding aqueous ammonia to solutions containing various metal ions
» Table 14.5 A variety of tests for aqueous anions
» Table 14.6 Tests for gases

Appearance or smell

A preliminary examination of the substance will give you a start. The appearance or smell of a substance can often indicate what it might contain (Table 14.1).

▼ **Table 14.1** Deductions that can be made from a substance's appearance or smell

Observation on substance	Indication
Black powder	Carbon, or contains O^{2-} ions (as in CuO), or S^{2-} ions (as in CuS)
Pale green crystals	Contains Fe^{2+} ions (as in iron(II) salts)
Dark green crystals	Contains Ni^{2+} ions (as in nickel(II) salts)
Blue or blue-green crystals	Contains Cu^{2+} ions (as in copper(II) salts)
Yellow-brown crystals	Contains Fe^{3+} ions (as in iron(III) salts)
Smell of ammonia	Contains NH_4^+ ions (as in ammonium salts)

Flame colours

If a wooden splint, which has been soaked in an aqueous metal ion solution, is held in a colourless Bunsen flame, the flame colour can become coloured (Figure 14.27). Certain metal ions may be detected in their compounds by observing their flame colours (Table 14.2).

▲ **Figure 14.27** The green colour is characteristic of copper

▼ **Table 14.2** Characteristic flame colours of some metal ions

	Metal	Flame colour
Group I (1+ ion)	Lithium	Red
	Sodium	Yellow
	Potassium	Lilac
Group II (2+ ion)	Calcium	Orange-red
	Barium	Light green
Others	Copper (as Cu^{2+})	Blue-green

A flame colour is obtained as a result of the electrons in the particular ions being excited when they absorb energy from the flame which is then emitted as visible light. The different electronic configurations of the different ions, therefore, give rise to the different colours.

Tests for aqueous cations

Effect of adding dilute sodium hydroxide solution

Aqueous sodium hydroxide can be used to identify salts of Al^{3+}, Ca^{2+}, Cr^{3+}, Cu^{2+}, Fe^{2+}, Fe^{3+} and Zn^{2+} when present in aqueous solutions. All metal cations form insoluble hydroxides when sodium hydroxide solution is added to them. The colour of the precipitate and its behaviour in excess sodium hydroxide solution will help identify the metal present (Table 14.3).

▼ **Table 14.3** Effect of adding sodium hydroxide solution to solutions containing various metal ions

Added dropwise	To excess	Likely cation
White precipitate	Precipitate is soluble in excess, giving a colourless solution	Al^{3+}, Zn^{2+}
White precipitate	Precipitate is insoluble in excess	Ca^{2+}
Green precipitate	Precipitate is insoluble in excess	Cr^{3+}
Light blue precipitate	Precipitate is insoluble in excess	Cu^{2+}
Green precipitate	Precipitate is insoluble in excess, turns brown near the surface on standing	Fe^{2+}
Red-brown precipitate	Precipitate is insoluble in excess	Fe^{3+}

In the case of ammonium salts containing the ammonium ion, NH_4^+, ammonia gas is produced on warming. The ammonium cation does not form an insoluble hydroxide. However, it forms ammonia and water upon heating.

▼ **Table** 14.4 Effect of adding aqueous ammonia to solutions containing various metal ions

Added dropwise	To excess	Cation present
White precipitate	Precipitate is insoluble in excess	Al^{3+}
No precipitate or very slight white precipitate	No change	Ca^{2+}
Green precipitate	Precipitate is insoluble in excess	Cr^{3+}
Light blue precipitate	Precipitate is soluble in excess, giving a dark blue solution	Cu^{2+}
Green precipitate	Precipitate is insoluble in excess, turns brown near the surface on standing	Fe^{2+}
Red-brown precipitate	Precipitate is insoluble in excess	Fe^{3+}
White precipitate	Precipitate is soluble in excess, giving a colourless solution	Zn^{2+}

▲ **Figure 14.28** Addition of sodium hydroxide will show the difference between Fe^{2+} and Fe^{3+} in aqueous solution

Effect of adding dilute ammonia solution

Ammonia gas dissolved in water is usually known as aqueous ammonia. The solution is only weakly alkaline, which results in a relatively low concentration of hydroxide ions. Aqueous ammonia can be used to identify salts of Al^{3+}, Ca^{2+}, Cr^{3+}, Cu^{2+}, Fe^{2+}, Fe^{3+} and Zn^{2+} ions. The colour of the precipitate or solution formed identifies the metal present (Table 14.4).

▲ **Figure 14.29** When aqueous ammonia is added to a solution containing Cu^{2+} ions, the solution forms a gelatinous light blue precipitate. As more is added to excess, the precipitate dissolves, forming a dark blue clear solution

Tests for aqueous anions

Table 14.5 shows a variety of tests for aqueous anions.

▼ **Table 14.5** Tests for aqueous anions

Anion	Test	Test result
Carbonate (CO_3^{2-})	Add dilute acid	Effervescence is seen as carbon dioxide produced
Chloride (Cl^-) [in solution]	Acidify with dilute nitric acid, then add aqueous silver nitrate	A white precipitate is produced
Bromide (Br^-) [in solution]	Acidify with dilute nitric acid, then add aqueous silver nitrate	A cream precipitate is produced
Iodide (I^-) [in solution]	Acidify with dilute nitric acid, then add aqueous silver nitrate	A yellow precipitate is produced
Nitrate (NO_3^-) [in solution]	Add aqueous sodium hydroxide, then aluminium foil; warm carefully	Ammonia gas is produced
Sulfate (SO_4^{2-}) [in solution]	Acidify, then add aqueous barium nitrate or barium chloride	A white precipitate is produced
Sulfite (SO_3^{2-})	Add a small volume of acidified aqueous potassium manganite(VIII)	The acidified aqueous potassium manganate(VII) changes colour from purple to colourless

▲ **Figure 14.30** The test for sulfate ions

Tests for gases

Table 14.6 shows the common gases which may be produced during qualitative analysis and tests which can be used to identify them. These tests are used in conjunction with the tests shown above.

▼ **Table 14.6** Tests for gases

Gas	Colour (odour)	Effect of moist indicator paper	Test
Hydrogen (H_2)	Colourless (odourless)	No effect – neutral	'Pops' in the presence of a lighted splint
Oxygen (O_2)	Colourless (odourless)	No effect – neutral	Relights a glowing splint
Carbon dioxide (CO_2)	Colourless (odourless)	Pink – weakly acidic	Turns limewater a cloudy white
Ammonia (NH_3)	Colourless (very pungent smell)	Blue – alkaline	Turns damp red litmus blue
Sulfur dioxide (SO_2)	Colourless (very choking smell)	Red – acidic	Turns acidified potassium dichromate(VI) from orange to green\n\nTurns acidified potassium manganate(VII) from purple to colourless
Chlorine (Cl_2)	Yellow-green (very choking smell)	Bleaches moist indicator paper after it initially turns pale pink	Bleaches damp litmus paper
Water (H_2O)	Colourless (odourless)	No effect – neutral	Turns blue cobalt chloride paper pink\n\nTurns anhydrous copper(II) sulfate from white to blue

▲ **Figure 14.31** Testing for oxygen gas

Practical skills

Testing a compound

For safe experiments/demonstrations which are related to this chapter, please refer to the *Cambridge IGCSE Chemistry Practical Skills Workbook*, which is also part of this series.

Safety

● Eye protection must be worn.

A student was given small quantities of a compound labelled A. After putting on their eye protection, the student carried out a variety of tests to help them identify compound A. Table 14.7 shows the tests the student did on compound A and the observations made.

Complete the table below by giving the possible conclusions to tests (a), (b), (c) and (d). Suggest both the test and observation that lead to the conclusion in test (d).

1 What is the name and formula for compound A?
2 It was suggested by the teacher that the final test (d) should be carried out in a fume cupboard. Explain why this procedure should be carried out in a fume cupboard.

▼ **Table 14.7** Tests on compound A and observations made

Test	Observation	Conclusion
(a) Compound A was dissolved in water and the solution produced divided into three parts for the tests (b), (c) and (d)	When dissolved in water, a colourless solution was produced.	
(b) (i) To the first sample, a small amount of aqueous sodium hydroxide was added. (ii) An excess of aqueous sodium hydroxide was added to the mixture from (i).	(i) White precipitate was produced. (ii) The white precipitate dissolved in excess sodium hydroxide.	
(c) (i) To the second part, a small amount of aqueous ammonia was added. (ii) An excess of aqueous ammonia was then added to the mixture from (i).	(i) A white precipitate was formed. (ii) The white precipitate was insoluble in excess.	
(d) Aqueous sodium hydroxide solution was added to some of the solution of A followed by some aluminium foil. The mixture was then warmed carefully.	A gas was produced which turned moist litmus paper blue.	

▶ Test yourself

4 What will be the colour of the precipitate formed with silver nitrate for the following?
   a chloride, Cl⁻   **c** iodide, I⁻
   b bromide, Br⁻
   Choose from yellow, white, green, cream, blue.
5 Match the gas shown in the table to the correct test for that gas and the positive result.

Gas	Test result
1 Oxygen	a Damp indicator paper turns white (bleached)
2 Carbon dioxide	b Burning splint produces a 'pop' sound when plunged into a test tube of gas
3 Hydrogen	c Damp litmus paper turns blue
4 Ammonia	d Limewater turns milky white
5 Chlorine	e Relights a glowing splint

Revision checklist

After studying Chapter 14 you should be able to:

✔ Name appropriate apparatus for the measurement of time, temperature, mass and volume.

✔ Suggest advantages and disadvantages of experimental methods and apparatus.

✔ Know the terms 'solvent', 'solute', 'solution', 'saturated solution', 'residue' and 'filtrate'.

✔ Describe an acid–base titration and know how to identify the end-point of a titration.

✔ Describe how paper chromatography is used to separate mixtures of soluble substances.

✔ Describe the use of locating agents when separating mixtures.

✔ Interpret simple chromatograms.

✔ State and use the equation:

$$R_f = \frac{\text{distance travelled by solute}}{\text{distance travelled by solvent}}$$

✔ Describe and explain methods of separation and purification.

✔ Suggest suitable separation and purification techniques.

✔ Identify substances and assess their purity using melting point and boiling point information.

✔ Describe tests using aqueous sodium hydroxide and aqueous ammonia to identify aqueous cations mentioned in this chapter in Tables 14.3 and 14.4.

✔ Describe the use of a flame test to identify the cations mentioned in this chapter in Table 14.2.

✔ Describe tests to identify the anions mentioned in this chapter in Table 14.5.

✔ Describe tests to identify the gases mentioned in this chapter in Table 14.6.

Exam-style questions

1 a Which of the following:
cm³, kilograms, hours, degrees Celsius, dm³, minutes, seconds, grams
are units of:
  i time [1]
  ii temperature [1]
  iii volume [1]
  iv mass? [1]

b In experiments, you will require the use of accurate measuring instruments. State the precision you would expect from the following measuring instruments:
  i a stopwatch [1]
  ii a thermometer [1]
  iii an electronic balance [1]
  iv a burette. [1]

2 State the method which is most suitable for separating the following:
  a oxygen from liquid air [1]
  b red blood cells from plasma [1]
  c petrol and kerosene from crude oil [1]
  d coffee grains from coffee solution [1]
  e pieces of steel from engine oil [1]
  f amino acids from fruit juice solution [1]
  g ethanol and water. [1]

3 Three solid dyes are to be separated. One is yellow, one is red and one blue. The following facts are known about the dyes.
  a The blue and yellow dyes are soluble in cold water.
  b The red dye is insoluble.
  c When chalk is added to the solution of blue and yellow dyes, the mixture stirred and then filtered, the filtrate is blue and the residue is yellow.
  d When the residue is mixed with alcohol and the mixture stirred and filtered, the filtrate is yellow and the residue is white.
Describe an experiment to obtain solid samples of each of the dyes from a mixture of all three. [10]

4 Forensic scientists recovered several samples from a crime scene. Chromatography was carried out on these samples. The scientists placed the samples onto chromatography paper. The chromatogram produced was sprayed with a *locating agent*.
The R_f *values* were then measured for each of the samples taken.

a What do you understand by the term:
  i locating agent? [2]
  ii R_f value? [2]
b The samples were thought to be amino acids. Use your research skills to find out the normal locating agent used for amino acids. [1]
c The table below shows R_f values for several amino acids thought to be at the crime scene and for three samples.

Amino acid	R_f value
Alanine	0.65
Glycine	0.25
Threonine	0.57
Proline	0.60
Lysine	0.16
Methionine	0.50
Sample 1	0.64
Sample 2	0.18
Sample 3	0.24

  i Identify the samples taken from the crime scene using the data from the table. [3]
  ii Explain how you made your choices in part **i**. [1]

5 For each of the following pairs of substances, describe a chemical test you would carry out to distinguish between them.
  a potassium sulfate and potassium sulfite [3]
  b ammonium chloride and aluminium chloride [3]
  c zinc nitrate and calcium nitrate [3]
  d sodium chloride and sodium iodide [3]
  e iron(II) sulfate and copper(II) sulfate [3]

6 Sodium carbonate hydrate contains water of crystallisation. When it is heated strongly, it gives off the water of crystallisation, which can be collected.
  a The substance left behind is anhydrous sodium carbonate. Describe a chemical test to show that this substance contains sodium (cation) and carbonate (anion). [2]
  b Describe two chemical tests to show that the colourless liquid produced and collected is indeed water. [4]
  c Describe another test to show that the colourless liquid given off in this experiment is pure water. [1]

1 Substances can be classified as elements, compounds or mixtures.
   a What is meant by the term *compound*? [2]
   b Mixtures can be separated by physical processes.
   A sequence of physical processes can be used to separate common salt (sodium chloride) from a mixture containing sand and common salt only.
   Give the order and the correct scientific term for the physical processes used to separate the common salt from the mixture.
   1
   2
   3 [4]
   The boiling points of four different alcohols, **A, B, C** and **D**, are shown.

alcohol	A	B	C	D
boiling point / °C	56	78	122	160

   c A student suggested that the apparatus shown could be used to separate the mixture of alcohols.

   i Apparatus **X** needs to have cold water flowing through it.
      – Draw an arrow on the diagram to show where the cold water enters apparatus **X**.
      – Name apparatus **X**. [2]
   ii Part of the fractionating column is missing. This means that the experiment will not work.
      – Draw on the diagram the part of the fractionating column which is missing.

   – Explain why the experiment will **not** work with this part of the fractionating column missing. [2]
   iii Suggest why a Bunsen burner is **not** used to heat the flask. [1]
   iv A hot water bath cannot be used to separate alcohols **C** and **D**.
      Explain why. [2]
   [Total: 13]

(Cambridge IGCSE Chemistry 0620 Paper 41 Q1 May/June 2018)

2 This question is about masses, volumes and moles.
   a Which term is defined by the following statement? [1]

      The average mass of naturally occurring atoms of an element on a scale where the ^{12}C atom has a mass of exactly 12 units.

   b Butane, C_4H_{10}, has a relative **molecular** mass of 58.
   Potassium fluoride, KF, has a relative **formula** mass of 58.
   Explain why the term relative molecular mass can be used for butane but **cannot** be used for potassium fluoride. [2]
   c A 0.095 g sample of gaseous element **Y** occupies 60.0 cm^3 at room temperature and pressure.
      – Determine the number of moles of element **Y** in 60.0 cm^3.
      moles of element **Y** = _____ mol
      – Calculate the relative molecular mass of element **Y** and hence suggest the identity of element **Y**.
      relative molecular mass = _____
      identity of element **Y** = _____ [3]
   d A 1.68 g sample of phosphorus was burned and formed 3.87 g of an oxide of phosphorus. Calculate the empirical formula of this oxide of phosphorus.
      empirical formula = _____ [4]
   e Another oxide of phosphorus has the empirical formula P_2O_3.

One molecule of this oxide of phosphorus contains four atoms of phosphorus.
Calculate the mass of **one** mole of this oxide of phosphorus.

mass = _____ g   [2]

[Total: 12]

(Cambridge IGCSE Chemistry 0620 Paper 41 Q4 May/June 2018)

3 a The table gives some chemical properties of transition elements and their compounds, and of Group I elements and their compounds.

chemical property	transition elements	Group I elements
ability to act as catalysts	yes	no
exist as coloured compounds	yes	no

  i What is meant by the term *catalyst*?   [2]
  ii Give **one** other chemical property shown by transition elements which is **not** shown by Group I elements.   [1]

b Give **two** physical properties shown by transition elements which are **not** shown by Group I elements.

  1 _____
  2 _____   [2]

c The energy level diagram shows the energy profile for the reaction between zinc and dilute sulfuric acid.

  i Complete the diagram by adding the formulae of the products. Include state symbols.   [3]
  ii Draw an arrow on the diagram to represent the activation energy.   [1]
  iii Is the reaction endothermic or exothermic? Explain your answer.   [1]

d The reaction between zinc and dilute sulfuric acid can be catalysed by the addition of aqueous copper(II) sulfate.
On the diagram, add the energy profile for the catalysed reaction.   [1]

e A student electrolyses aqueous copper(II) sulfate using the apparatus shown.

Oxygen gas forms at the positive electrode (anode).
  i Write an ionic half-equation for the reaction at the negative electrode (cathode). Include state symbols.   [3]
  ii Describe what the student observes at the negative electrode.   [1]
  iii Give **two other** observations which the student makes during the electrolysis.
  1
  2   [2]
  iv What difference would the student observe at the positive electrode if the aqueous copper(II) sulfate were replaced by concentrated aqueous copper(II) chloride?   [1]

[Total: 18]

(Cambridge IGCSE Chemistry 0620 Paper 41 Q5 May/June 2018)

4 The table shows the structures of four hydrocarbons.

P	Q	R	S
CH_3-CH_3	$CH_2=CH_2$	$CH_2=CH-CH_3$	$CH_2=CH-CH_2-CH_3$

a Why are compounds **P**, **Q**, **R** and **S** known as hydrocarbons?   [2]
b Compound **P** is saturated. What is meant by the term *saturated*?   [1]

c Compound **P** undergoes a substitution reaction with chlorine.
   i What is meant by the term *substitution reaction*? [1]
   ii State a condition required for this reaction to occur. [1]
   iii Write a chemical equation for this reaction. [2]
d Compound **R** undergoes an addition reaction with bromine.
   i Why is this reaction an addition reaction? [1]
   ii A compound containing bromine is formed in this reaction.
      Draw the structure of this compound. Show all of the atoms and all of the bonds. [1]
e Draw the structure of an unbranched isomer of compound **S**. Show all of the atoms and all of the bonds. Name this unbranched isomer of compound **S**. [2]
f Compound **Q** undergoes polymerisation.
   i Name the polymer formed. [1]
   ii Complete the chemical equation to show the polymerisation of compound **Q**.
      $n\ CH_2=CH_2 \rightarrow$ [2]
g Amino acids undergo polymerisation to form proteins. Part of a protein molecule with the linkages missing is shown.
   Draw the linkages on the diagram. Show all of the atoms and all of the bonds. [2]

h The structure shows an ester.

Write the word equation for a reaction which could be used to make this ester. [3]
[Total: 19]

(Cambridge IGCSE Chemistry 0620 Paper 41 Q6 May/June 2018)

5 The table shows the percentage by volume of each of the gases present in the exhaust gases from a petrol engine.

name	percentage by volume
carbon monoxide	1.0
carbon dioxide	
hydrogen	0.2
nitrogen	77.0
nitrogen dioxide	0.3
oxygen	0.7
hydrocarbons	0.3
water vapour	5.0
	total 100.0

a i Calculate the percentage by volume of carbon dioxide in the exhaust gases.
     _____% [1]
  ii Which gas shown in the table is present in the lowest percentage by volume? [1]
  iii Which **two** elements in the table combine to form nitrogen dioxide? [1]
  iv Give the formula for nitrogen dioxide. [1]
  v Where does the nitrogen in the exhaust gases come from? [1]
b The carbon monoxide in the exhaust gases comes from the incomplete combustion of hydrocarbons.
  i What is meant by the term *hydrocarbon*? [2]
  ii Give **one** adverse effect of carbon monoxide on health. [1]
  iii Balance the chemical equation for the complete combustion of pentane.

$$C_5H_{12} + 8O_2 \rightarrow .... CO_2 + ... H_2O$$ [2]
[Total: 10]

(Cambridge IGCSE Chemistry 0620 Paper 31 Q2 May/June 2018)

6 This question is about iron and its compounds.
  a The table shows how easy it is to reduce four metal oxides by heating with carbon.

metal oxide	ease of reduction with carbon
chromium(III) oxide	only reduced above 1700°C
iron(III) oxide	only reduced above 650°C
magnesium oxide	**not** reduced at 1750°C
nickel(II) oxide	only reduced above 300°C

Use this information to put the metals in order of their reactivity. Put the least reactive metal first. [2]

least reactive ──────────────► most reactive

b Iron is a transition element. Potassium is an element in Group I of the Periodic Table. Describe **three** ways in which the properties of iron differ from those of potassium. [3]

c Iron wire burns in oxygen. Balance the chemical equation for this reaction. [2]

..... Fe + ... $O_2 \rightarrow Fe_3O_4$

d Pure iron can be made by reducing iron(III) oxide, Fe_2O_3, with hydrogen.

$Fe_2O_3 + 3H_2 \rightarrow 2Fe + 3H_2O$

How does this equation show that iron(III) oxide is reduced? [1]

e When iron reacts with dilute hydrochloric acid, iron(II) chloride is formed.
   i Describe a test for iron(II) ions.

   test _____

   result _____ [2]

   ii Another chloride of iron has the structure shown.

   Deduce the molecular formula of this compound showing the number of iron and chlorine atoms. [1]

f Some iron nails were placed in bottles under different conditions.

In which bottles will the iron nails **not** rust? Give reasons for your answer. [2]

[Total: 13]

(Cambridge IGCSE Chemistry 0620 Paper 31 Q4 May/June 2018)

7 a Complete the sentence about electrolysis using words from the list.

breakdown   compound   electricity electroplating   element   gaseous heat   molten

Electrolysis is the _____ of an ionic _____ when _____ or in aqueous solution by the passage of _____ [4]

b Molten zinc iodide can be electrolysed using the apparatus shown.

On the diagram, label:
   – the anode
   – the cathode
   – the electrolyte [2]

c Why are the electrodes made of graphite? [1]

d Predict the products of the electrolysis of molten zinc iodide at:
   the negative electrode
   the positive electrode. [2]

e When chlorine is bubbled through a colourless aqueous solution of zinc iodide, the solution turns brown.
   Name the brown substance. Suggest, using ideas about reactivity of the halogens, why this reaction occurs. [2]

[Total: 11]

(Cambridge IGCSE Chemistry 0620 Paper 31 Q5 May/June 2018)

8 This question is about isotopes.
   a An atom of an isotope of fluorine is represented by the symbol shown.

   $^{19}_{9}F$

   Describe the structure of an atom of this isotope of fluorine.

PAST PAPER QUESTIONS

. . . .

In your answer, include:
- the position of the protons, neutrons and electrons in the atom
- the number of protons, neutrons and electrons present in the atom. [5]

b Complete the sentence about isotopes using words from the list.

atomic   compound   element   ions   molecular   nucleons

Isotopes are atoms of the same _____

which have the same _____ number

but different numbers of _____ . [3]

c Give **one** medical use of radioactive isotopes. [1]

d Which **one** of the following isotopes is used as a source of energy?
   Draw a circle around the correct answer. [1]

$^{127}_{53}I$ $\quad$ $^{235}_{92}U$ $\quad$ $^{131}_{54}Xe$ $\quad$ $^{66}_{30}Z$

[Total: 10]

(Cambridge IGCSE Chemistry 0620 Paper 31 Q6 May/June 2018)

9 When zinc reacts with hydrochloric acid, hydrogen gas is produced.
   The graph shows how the volume of hydrogen gas produced changes with time when an excess of zinc is reacted with 0.2 mol/dm³ hydrochloric acid.

volume of hydrogen gas/cm³

time/min

a Explain why the volume of hydrogen gas remains constant after six minutes. [1]

b What volume of hydrogen gas was released in the first **two** minutes of the reaction? [1]

c The experiment is repeated using the same volume of 0.1 mol/dm³ hydrochloric acid. All other conditions are kept the same.
   On the grid, draw the graph for the experiment using 0.1 mol/dm³ hydrochloric acid. [2]

d Give the name of the salt formed when zinc reacts with hydrochloric acid. [1]

e Which **one** of the following pH values could be the pH of dilute hydrochloric acid? Draw a circle around the correct answer.
   pH 1 $\quad$ pH 7 $\quad$ pH 9 $\quad$ pH 13 [1]

[Total: 6]

(Cambridge IGCSE Chemistry 0620 Paper 31 Q8 May/June 2018)

10 This question is about subatomic particles.
    a Define the terms:
       proton number
       nucleon number [3]
    b Why is the ^1_1H hydrogen atom the **only** atom to have an identical proton number and nucleon number? [1]
    c Complete the table to show the number of protons, neutrons and electrons in the atoms and ions given. [6]

	number of protons	number of neutrons	number of electrons
^{19}F			9
^{26}Mg	12		
$^{31}P^{3-}$			
$^{87}Sr^{2+}$			

d i Write the formula of the compound formed from fluorine and magnesium. [1]
   ii Write the formula of the compound formed from Sr^{2+} and P^{3-}. [1]

[Total: 12]

(Cambridge IGCSE Chemistry 0620 Paper 41 Q1 May/June 2017)

11 Zinc is a very important metal.
    a When rods of zinc and copper are placed into dilute sulfuric acid as shown, electricity is generated.

bulb

zinc rod

copper rod

dilute sulfuric acid

i Write the ionic half-equation for the reaction occurring at the zinc rod. [2]

ii Write the ionic half-equation for the reaction occurring at the copper rod. [2]

iii The copper rod was replaced by an iron rod. Suggest the change, if any, in the intensity of the light emitted from the bulb and give a reason for your answer.

change:

reason: [2]

[Total: 6]

(Cambridge IGCSE Chemistry 0620 Paper 41 Q4c May/June 2017)

12 The alkenes and alkanes are both examples of homologous series which are hydrocarbons.

a What is meant by the term *hydrocarbon*? [2]

b Give **three** characteristics of an homologous series.

1 _____

2 _____

3 _____ [3]

c Name and draw the structure of the second member of the alkene homologous series. Show all of the atoms and all of the bonds.

name

structure [2]

d Alcohols can be made from alkenes. Name the reagent and conditions needed to convert an alkene into an alcohol. [2]

e The alcohol butanol, $CH_3CH_2CH_2CH_2OH$, can be converted into a carboxylic acid with four carbon atoms.

i Name the carboxylic acid formed from butanol and draw its structure. Show all of the atoms and all of the bonds.

name

structure [2]

ii Ethanoic acid can be formed from ethanol by fermentation. It can also be formed by the addition of a suitable chemical reagent. Name the reagent needed to convert ethanol into ethanoic acid. [2]

iii State the type of chemical change which occurs when ethanol is converted into ethanoic acid. [1]

f Describe how a student could prepare the ester methyl ethanoate in a school laboratory. In your description give

– the names of the **two** starting organic chemicals,

– the essential reaction conditions needed,

– a chemical equation for the reaction. [5]

[Total: 19]

(Cambridge IGCSE Chemistry 0620 Paper 41 Q6 May/June 2017)

13 a The electronic structures of five atoms, **A**, **B**, **C**, **D** and **E**, are shown.

    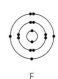

A   B   C   D   E

Answer the following questions about these atoms. Each atom may be used once, more than once or not at all.

Which atom, **A**, **B**, **C**, **D** or **E**,

i has a complete outer shell of electrons [1]

ii has a proton number of 15 [1]

iii has 4 shells containing electrons [1]

iv is a fluorine atom [1]

v is an atom of a metallic element? [1]

b Complete the table to show the number of electrons, neutrons and protons in the chlorine atom and bromide ion shown.

	number of electrons	number of neutrons	number of protons
$^{35}_{17}Cl$	17		
$^{79}_{35}Br^-$		44	

[3]

[Total: 8]

(Cambridge IGCSE Chemistry 0620 Paper 31 Q1 May/June 2017)

14 a The table shows the ions present in a 1000 cm³ sample of polluted river water.

ion present	formula of ion	mass present in mg / 1000 cm³
calcium	Ca^{2+}	2.0
chloride	Cl^-	1.3
hydrogencarbonate	HCO_3^-	2.0
magnesium	Mg^{2+}	1.0
potassium	K^+	4.0
silicate	SiO_3^{2-}	12.0
sodium	Na^+	11.0
	SO_3^{2-}	3.0

Answer these questions using the information from the table.

   i Which positive ion is present in the highest concentration? [1]

   ii State the name of the ion SO_3^{2-}. [1]

   iii Calculate the mass of silicate ions present in 250 cm³ of this sample.

      mass of silicate ions = _____ mg [1]

   iv Calculate the mass of solid formed when all the water is evaporated from the 1000 cm³ sample.

      mass of solid formed = _____ mg [1]

   v Name the compound containing Ca^{2+} ions and HCO_3^- ions. [1]

  b Describe a test for sodium ions.

    test _____

    result _____ [2]

  c The formulae of some chlorides are given.
    aluminium chloride, $AlCl_3$
    calcium chloride, $CaCl_2$
    sodium chloride, NaCl
    Deduce the formula for magnesium chloride. [1]

  d Molten calcium chloride can be electrolysed using inert electrodes.
    Predict the products of this electrolysis at

    the negative electrode (cathode) _____

    the positive electrode (anode) _____ [2]

                           [Total: 10]

(Cambridge IGCSE Chemistry 0620 Paper 31 Q2 May/June 2017)

15 The diagram shows part of the structures of three substances, **X**, **Y** and **Z**, at room temperature and pressure

        X   Y   Z

  a Describe substances **X**, **Y** and **Z** in terms of
    – their bonding,
    – the arrangement of their particles,
    – the motion of their particles. [5]

  b A closed gas syringe contains substance **X**.

Describe what happens to the volume of substance **X** in the syringe when the pressure is increased. The temperature remains constant. Explain your answer in terms of particles. [2]

  c Substance **Z** is diamond. Diamond is used in jewellery.
    Give **one** other use of diamond. [1]

  d Substance **Y** undergoes physical and chemical changes.
    Which **two** of the following are physical changes? Explain your answer.
    A Substance **Y** dissolves easily in water.
    B An aqueous solution of substance **Y** gives a white precipitate with acidified aqueous silver nitrate.
    C Substance **Y** melts at 801°C.
    D Substance **Y** reacts with concentrated sulfuric acid. [3]

                           [Total: 11]

(Cambridge IGCSE Chemistry 0620 Paper 31 Q3 May/June 2017)

16 Chlorine is an element in Group VII of the Periodic Table.

  a Chlorine is a diatomic molecule.
    Explain what is meant by the term *diatomic*. [1]

  b Aqueous chlorine reacts with aqueous potassium bromide.

$$Cl_2 + 2KBr \rightarrow Br_2 + 2KCl$$

   i How does this equation show that chlorine is more reactive than bromine? [1]

   ii Aqueous potassium bromide and aqueous potassium chloride are both colourless. Predict the colour change when aqueous chlorine reacts with aqueous potassium bromide. [1]

   iii Complete the chemical equation for the reaction of aqueous bromine with aqueous potassium iodide.

$$Br_2 + 2KI \rightarrow \text{.......} + \text{........}$$ [2]

  c Describe a test for iodide ions.

    test _____

    result _____ [2]

  d Chlorine is used to make the polymer PVC.

   i The monomer used to make PVC is made by the thermal decomposition of dichloroethane.

$$CH_2Cl\text{–}CH_2Cl \rightarrow CH_2{=}CHCl + HCl$$

Explain what is meant by the term *thermal decomposition*. [2]
ii PVC is a non-biodegradable plastic. Describe **two** pollution problems caused by non-biodegradable plastics.

1 _____

2 _____ [2]

[Total: 11]

(Cambridge IGCSE Chemistry 0620 Paper 31 Q4a, b, c, dii and iii May/June 2017)

17 Protons, neutrons and electrons are subatomic particles.
  a Complete the table to show the relative mass and relative charge of a proton, a neutron and an electron. [3]

particle	relative mass	relative charge
proton		
neutron		
electron	$\frac{1}{1840}$	

  b Bromine has two isotopes.
    i Define the term *isotope*. [2]
    ii Explain why the two isotopes of bromine have the same chemical properties. [2]
  c The table shows the number of protons, neutrons and electrons in some atoms and ions. Complete the table.

particle	number of protons	number of neutrons	number of electrons
$^{7}_{3}Li$			
$^{34}_{16}S^{-2}$			
	19	22	18

[5]

[Total: 12]

(Cambridge IGCSE Chemistry 0620 Paper 41 Q1 May/June 2016)

18 Period 3 contains the elements sodium to argon. This question asks about the chemistry of each of the Period 3 elements or their compounds.
  a Sodium nitrate is a white crystalline solid. When heated it melts and the following reaction occurs.

$$2NaNO_3(l) \rightarrow 2NaNO_2(l) + O_2(g)$$

A 3.40 g sample of sodium nitrate is heated. Calculate the
  – number of moles of $NaNO_3$ used,
      _____ mol
  – number of moles of O_2 formed,
      _____ mol
  – volume of O_2 formed, in dm^3 (measured at r.t.p.) _____ dm^3 [3]
  b Magnesium reacts slowly with warm water to form a base, magnesium hydroxide.
    i Explain what is meant by the term *base*. [1]
    ii Write a chemical equation for the reaction between magnesium and warm water. [2]
  c Aluminium oxide is amphoteric. It is insoluble in water. Describe experiments to show that aluminium oxide is amphoteric. [3]
  d Silicon(IV) oxide has a giant structure.
    i Name the type of bonding in silicon(IV) oxide. [1]
    ii Give two **physical** properties of silicon(IV) oxide. [2]
  e Calcium phosphate is used in fertilisers. The bonding in calcium phosphate is ionic. Calcium phosphate contains the phosphate ion, PO_4^{3-}.
    i What is ionic bonding? [2]
    ii Deduce the formula of calcium phosphate. [1]
  f Sulfur tetrafluoride, SF_4, can be made by combining gaseous sulfur with fluorine.

$$S(g) + 2F_2(g) \rightarrow SF_4(g)$$

The reaction is exothermic.
    i Complete the energy level diagram for this reaction. Include an arrow which clearly shows the energy change during the reaction.

[3]
    ii During the reaction the amount of energy given out is 780 kJ / mol.

The F–F bond energy is 160 kJ / mol. Use this information to determine the bond energy, in kJ / mol, of one S–F bond in SF_4.

$$S + \begin{array}{c} F—F \\ F—F \end{array} \rightarrow \begin{array}{c} F \\ | \\ F—S—F \\ | \\ F \end{array}$$

_____ kJ / mol   [3]

g Chlorine and compounds of chlorine are important in water treatment and in laboratory testing for water.

   i Chlorine is added to water to make the water safe to drink.
   Explain why adding chlorine makes water safe to drink.   [1]

   ii A compound of chlorine is used in the laboratory to test for the presence of water.
   Name the compound of chlorine used in this test and describe the colour change seen in a positive result of this test.

   name of compound _____

   colour change from _____

   to _____   [3]

h Argon is an unreactive noble gas.
Explain why argon is unreactive.   [1]

[Total: 26]

(Cambridge IGCSE Chemistry 0620 Paper 41 Q2a to hi May/June 2016)

19 Electroplating steel objects with silver involves a three-step process.

step 1 A coating of copper is applied to the object.
step 2 A coating of nickel is applied to the object.
step 3 The coating of silver is applied to the object.

a A diagram of the apparatus used for **step 1** is shown.

copper (anode)

steel object (cathode)

aqueous copper(II) sulfate electrolyte

   i The chemical process taking place on the surface of the object is
   $$Cu^{2+}(aq) + 2e^- \rightarrow Cu(s)$$
   Explain whether this process is oxidation or reduction.   [1]

   ii Explain why the concentration of copper ions in the electrolyte remains constant throughout **step 1**.   [2]

b Give **two** changes which would be needed in order to coat nickel onto the object in **step 2**.   [2]

c Copper, nickel and silver are transition elements. Typical physical properties of transition elements are a high density and a high melting point.
Give **three** different properties of transition metals which are not typical of other metals.[3]

[Total: 8]

(Cambridge IGCSE Chemistry 0620 Paper 41 Q4 May/June 2016)

20 Petroleum is a source of many important chemicals.

a Name **two** industrial processes which must take place to produce alkenes from petroleum.   [2]

b Ethene, $CH_2=CH_2$, and propene, $CH_2=CHCH_3$, can both be converted into polymers.

   i What type of polymerisation takes place when ethene forms a polymer?   [1]

   ii What is the empirical formula of the polymer formed from ethene?   [1]

   iii Propene has the structural formula $CH_2=CHCH_3$. Draw **two** repeat units of the polymer made from propene.   [2]

c Ethene will react with steam to form ethanol. Propene will react with steam to form two isomers, both of which are alcohols.
Suggest the structures of these alcohols.   [2]

d Esters are organic chemicals noted for their characteristic smells. Ethanoic acid and methanol will react to form an ester.

   i Name the catalyst needed to form an ester from ethanoic acid and methanol.   [1]

   ii Name the ester formed when ethanoic acid reacts with methanol.   [1]

   iii Draw the structure of the ester formed when ethanoic acid reacts with methanol. Show all bonds.   [2]

   iv Give the name of a polyester.   [1]

[Total: 13]

(Cambridge IGCSE Chemistry 0620 Paper 41 Q6 May/June 2016)

21 The structures of some substances containing chlorine are shown.

a Answer the following questions about these substances.
  i Which substance is a diatomic molecule? [1]
  ii Which substance represents part of an ionic structure? [1]
  iii Which substance is an element? Explain your answer. [2]
  iv Determine the simplest formula for substance **D**. [1]

b The symbols for two isotopes of chlorine are shown.

$$^{35}_{17}Cl \qquad ^{37}_{17}Cl$$

  i How do these two isotopes differ in their atomic structure? [1]
  ii Determine the number of neutrons present in one atom of the isotope $^{35}_{17}Cl$ [1]
  iii Draw the electronic structure of a chlorine atom. Show all shells and all electrons. [2]

[Total: 9]

(Cambridge IGCSE Chemistry 0620 Paper 31 Q1 May/June 2016)

22 A bicycle maker wants to choose a suitable material to make bicycle frames.
The table shows the properties of some materials that could be used.

material	relative strength	density in g /cm³	resistance to corrosion	cost per tonne in $ / tonne
aluminium	8	2.7	very good	1500
iron	21	7.9	poor	450
stainless steel	24	7.9	very good	600
titanium	27	4.5	very good	15000
zinc	14	7.1	good	1300

a Which material is the most suitable for making the bicycle frame?
Explain your answer using information from the table. [3]
b Aluminium is extracted from aluminium oxide by electrolysis.
  i State the name of the main ore of aluminium. [1]
  ii Suggest why aluminium is extracted by electrolysis and **not** by reduction with carbon. [1]
  iii Molten aluminium oxide is electrolysed using graphite electrodes.
  Predict the products of this electrolysis at the positive electrode (anode),

  _____

  the negative electrode (cathode).

  _____ [2]

[Total: 7]

(Cambridge IGCSE Chemistry 0620 Paper 31 Q2a and b May/June 2016)

23 The table shows some properties of the Group I metals.

metal	density in g /cm³	melting point / °C	boiling point / °C
lithium	0.53	181	1342
sodium		98	883
potassium	0.86	63	760
rubidium	1.53	39	686
caesium		29	669

a i Describe the trend in boiling points of the Group I metals. [1]
  ii Predict the density of caesium. [1]
  iii Deduce the state of caesium at 20°C. Explain your answer. [2]
b Complete the word equation for the reaction of rubidium with water.
rubidium + water → .... + ... [2]
c The dye, indigotin, is formed when compound **F** is exposed to air.
The structure of compound **F** is shown below.

Complete the table and calculate the relative molecular mass of compound **F**.

type of atom	number of atoms	atomic mass	
carbon	8	12	8 × 12 = 96
hydrogen			
nitrogen	1	14	1 × 14 = 14
oxygen	1	16	1 × 16 = 16
sodium			

relative molecular mass _____ [2]

d Three dye mixtures, **J**, **K** and **L**, were spotted onto a piece of chromatography paper. Three pure dyes, **X**, **Y** and **Z**, were also spotted onto the same piece of paper.
The diagram shows the results of this chromatography.

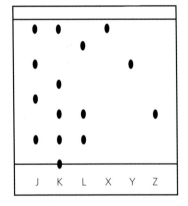

i Suggest why the base line was drawn in pencil and **not** in ink. [1]
ii Which dye mixture, **J**, **K** or **L**, contains a dye which did **not** move during this chromatography? [1]
iii Which dye mixture, **J**, **K** or **L**, contains both dye **X** and dye **Y**? [1]
iv Which dye mixture, **J**, **K** or **L**, does **not** contain dye **Z**? [1]

[Total: 12]

(Cambridge IGCSE Chemistry 0620 Paper 31 Q3 May/June 2016)

24 The structures of some organic compounds are shown.

a i Which **two** of these compounds are alcohols? Explain your answer. [2]
ii Which **two** of these compounds are saturated hydrocarbons? [1]
b Methanol and ethanol are alcohols in the same homologous series.
Complete the following sentence about a homologous series using words from the list.
alcohols chemical compounds elements functional mixtures physical
A homologous series is a family of similar _____ with similar _____ properties due to the same _____ group. [3]
c Ethene is an alkene.
i Draw the structure of ethene showing all atoms and all bonds. [1]
ii Describe how aqueous bromine is used to show that ethene is an unsaturated compound. [2]
iii Ethene is manufactured by cracking. State the conditions needed for cracking. [1]
iv Complete the chemical equation for the cracking of hexadecane, $C_{16}H_{34}$, to form propene and one other hydrocarbon. [1]

$$C_{16}H_{34} \rightarrow C_3H_6 + \text{.............}$$

[Total: 11]

(Cambridge IGCSE Chemistry 0620 Paper 31 Q4 May/June 2016)

25 Ammonia is manufactured by the reaction of nitrogen with hydrogen in the presence of a catalyst.

a What is the purpose of a catalyst? [1]

b The reaction is reversible.
Complete the equation below by adding the sign for a reversible reaction. [1]

$N_2 + 3H_2 \qquad 2NH_3$

c The energy level diagram for this reaction is shown.
Is this reaction exothermic or endothermic? Give a reason for your answer. [1]

d The graph shows how the percentage yield of ammonia changes with temperature when the pressure is kept constant.

i Describe how the percentage yield of ammonia changes with temperature. [1]

ii Determine the percentage yield of ammonia at 350°C. [1]

e Describe a test for ammonia.

test _____

result _____ [2]

f Ammonia is a weak base.
Describe how you would measure the pH of an aqueous solution of a weak base using Universal Indicator. [2]

g Complete the chemical equation for the reaction of ammonia with chlorine. [2]

$........\ NH_3 + 3Cl_2 \rightarrow N_2 + ...HCl$

[Total: 11]

(Cambridge IGCSE Chemistry 0620 Paper 31 Q6 May/ June 2016)

26 The diagrams show part of the structures of five substances, **A**, **B**, **C**, **D** and **E**.

A   B   C

D   E

a Answer the following questions about these structures.
Each structure may be used once, more than once or not at all.

i Which **two** of these structures, **A**, **B**, **C**, **D** or **E**, are covalently bonded?
_____ and _____ [2]

ii Which **one** of these structures, **A**, **B**, **C**, **D** or **E**, is a diatomic molecule? [1]

iii Which **one** of these structures, **A**, **B**, **C**, **D** or **E**, is a compound? [1]

iv Which **one** of these structures, **A**, **B**, **C**, **D** or **E**, is very soluble in water? [1]

v Which **one** of these structures, **A**, **B**, **C**, **D** or **E**, is used in cutting tools? [1]

vi Which **one** of these structures, **A**, **B**, **C**, **D** or **E**, is used in electrical wiring? [1]

b Substance **B** is an element.
What is meant by the term *element*? [1]

[Total: 8]

(Cambridge IGCSE Chemistry 0620 Paper 31 Q1 May/ June 2019)

27 This question is about iron and iron compounds.

a Name the main ore of iron. [1]

b In a blast furnace used for the extraction of iron, carbon reacts with oxygen from the air to form carbon monoxide.

Complete the chemical equation for this reaction. [2]

....C + ... → 2CO

c In the hotter parts of the furnace, carbon reacts with the iron(III) oxide present in the iron ore.

$3C + Fe_2O_3 \rightarrow 3CO + 2Fe$

How does this equation show that carbon is oxidised? [1]

d Limestone is added to the blast furnace. The limestone is converted into calcium oxide and carbon dioxide. The reaction is endothermic.

$CaCO_3 \xrightarrow{heat} CaO + CO_2$

i What type of chemical reaction is this? [1]

ii What type of oxide is calcium oxide? Give a reason for your answer. [2]

e Iron is a metal.
Give **three** physical properties that are characteristic of metals. [3]

f The structure of a compound of iron is shown.

Deduce the molecular formula of this compound to show the number of iron, carbon and oxygen atoms. [1]

[Total: 11]

(Cambridge IGCSE Chemistry 0620 Paper 31 Q2 May/ June 2019)

28 This question is about iodine and compounds of iodine.

a Use the kinetic particle model to describe the separation between the molecules and the type of motion of the molecules in:
• solid iodine
• iodine gas. [4]

b The graph shows how the volume of iodine gas changes with pressure. The temperature is kept constant.

Describe how the volume of iodine gas changes with pressure. [1]

c i Complete the word equation to show the halogen and halide compound which react to form the products iodine and potassium bromide. [2]

ii Explain, in terms of the reactivity of the halogens, why aqueous iodine does **not** react with aqueous potassium chloride. [1]

d Iodine reacts with aqueous sodium thiosulfate, $Na_2S_2O_3$.

i Balance the chemical equation for this reaction. [2]

.....$Na_2S_2O_3 + I_2 \rightarrow Na_2S_4O_6 + $.....$NaI$

ii The energy level diagram for this reaction is shown.

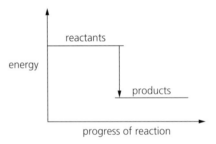

Explain how this diagram shows that the reaction is exothermic. [1]

e Describe a test for iodide ions.

test _____

observations _____ [2]

f Molten sodium iodide is electrolysed. Predict the product at the positive electrode. [1]

[Total: 14]

(Cambridge IGCSE Chemistry 0620 Paper 31 Q4 May/ June 2019)

29 This question is about copper and copper compounds.

a Describe how you could prepare a pure sample of crystals of hydrated copper(II) sulfate using dilute sulfuric acid and an excess of copper(II) oxide. [3]

b Anhydrous copper(II) sulfate is used to test for water.

$$CuSO_4 \quad + \quad 5H_2O \quad \rightleftharpoons \quad CuSO_4.5H_2O$$

anhydrous   hydrated
copper(II) sulfate   copper(II)sulfate

   i   What is meant by the symbol $\rightleftharpoons$?   [1]
   ii   How can hydrated copper(II) sulfate be changed into anhydrous copper(II) sulfate?   [1]
  c   Complete the table to calculate the relative formula mass of anhydrous copper(II) sulfate, $CuSO_4$.
     Use your Periodic Table to help you.

type of atom	number of atoms	relative atomic mass	
copper	1	64	$1 \times 64 = 64$
sulfur			
oxygen			

     relative formula mass = _____ [2]
  d   Complete the table to show the number of electrons, protons and neutrons in the sulfur atom and copper ion shown.

	number of electrons	number of neutrons	number of protons
$^{34}_{16}S$			
$^{63}_{29}Cu^{+2}$			29

                                                    [4]
  e   Alloys of copper are used to make coins.
   i   What is meant by the term *alloy*?   [1]
   ii   Suggest why an alloy of copper is used to make coins instead of using pure copper. [1]
                                                  [Total: 13]
(Cambridge IGCSE Chemistry 0620 Paper 31 Q6 May/
June 2019)

30 Ethanoic acid is a weak acid and hydrochloric acid is a strong acid.
Both ethanoic acid and hydrochloric acid dissociate in aqueous solution.
  a i   Define the term *acid*.   [1]
    ii   The chemical equation shows the changes which occur when the *strong* acid, hydrochloric acid, is added to water.

$$HCl\,(aq) \rightarrow H^+(aq) + Cl^-(aq)$$

        Complete the chemical equation to show the changes which occur when the *weak* acid, ethanoic acid, is added to water.

$$CH_3COOH(aq) _____ \quad [2]$$

b   A student does experiments to show that hydrochloric acid is a strong acid and ethanoic acid is a weak acid. The student adds an excess of hydrochloric acid and an excess of ethanoic acid to separate samples of lumps of calcium carbonate.
Only the identity of the acid is changed between the experiments. All other conditions are kept the same.
   i   State **two** observations which would show that hydrochloric acid is a stronger acid than ethanoic acid.   [2]
   ii   The student uses the same size container and checks that the pressure is the same for each experiment.
       State **three** other conditions which must be kept the same to ensure fair testing.[3]
c   Hydrochloric acid produces salts called chlorides. Magnesium carbonate reacts with hydrochloric acid to produce magnesium chloride.

$$MgCO_3 + 2HCl \rightarrow MgCl_2 + H_2O + CO_2$$

A student used 50.00 cm^3 of 2.00 mol/dm^3 hydrochloric acid in an experiment to produce magnesium chloride.
Calculate the mass, in g, of magnesium carbonate needed to react exactly with 50.00 cm^3 of 2.00 mol/dm^3 hydrochloric acid using the following steps.
•   Calculate the number of moles of HCl present in 50.00 cm^3 of 2.00 mol/dm^3 HCl.
       _____ mol
•   Determine the number of moles of $MgCO_3$ which would react with 50.00 cm^3 of 2.00 mol/dm^3 HCl.
       _____ mol
•   Calculate the relative formula mass, M_r, of $MgCO_3$.
       M_r of $MgCO_3$ = _____
•   Calculate the mass of $MgCO_3$ needed to react exactly with 50.00 cm^3 of 2.00 mol/dm^3 HCl.
       mass = _____ g   [4]
d   A student prepares crystals of magnesium chloride by adding an excess of magnesium carbonate to 50.00 cm^3 of 2.00 mol/dm^3 hydrochloric acid.
The student filters the mixture and rinses the residue.

  i Why does the student add an **excess** of magnesium carbonate? [1]
  ii Why does the student rinse the residue? [1]
  iii Describe how the student would obtain pure crystals of magnesium chloride from the filtrate. [3]
 e Silver chloride, AgCl, is insoluble. It can be made by a precipitation reaction between aqueous barium chloride and a suitable aqueous silver salt.
  i What is meant by the term *precipitate*? [2]
  ii Name a suitable silver salt to use to prepare silver chloride.
     Complete the chemical equation to show the formation of insoluble silver chloride from aqueous barium chloride and the silver salt you have named.

     name of a suitable silver salt _____

     $BaCl_2$ + ... → .... + ... [3]
     [Total: 22]
(Cambridge IGCSE Chemistry 0620 Paper 41 Q4 May/June 2019)

31 The structures of five alkenes, **A**, **B**, **C**, **D** and **E**, are shown.

 a What is the general formula of alkenes? [1]
 b What is the molecular formula of alkene **D**? [1]
 c Predict which alkene, **A**, **B**, **C**, **D** or **E**, has the highest boiling point.
   Explain your answer.
   alkene _____
   explanation _____ [2]
 d Which alkene, **A**, **B**, **C**, **D** or **E**, diffuses most quickly?
   Explain your answer.
   alkene _____
   explanation _____ [2]

 e A student added aqueous bromine to alkene **C**.

   Describe the colour change seen and draw the structure of the product. Show all of the atoms and all of the bonds.
   colour change from _____ to
   _____
   structure [2]
 f Two different alcohols can be produced from alkene **B** by an addition reaction.

   i Draw the structures of the **two** alcohols. Show all of the atoms and all of the bonds. [2]
   ii State the reagent and conditions needed to produce an alcohol from alkene **B**.
      reagent _____
      conditions _____ [3]
 g Alkene **C** can be converted into a polymer.

   i What type of polymerisation occurs? [1]
   ii Suggest the name of the polymer formed. [1]
   iii Complete the chemical equation to show this polymerisation. [3]

   iv State the empirical formula of the polymer formed. [1]
   [Total: 19]
(Cambridge IGCSE Chemistry 0620 Paper 41 Q5 May/June 2019)

Alternative to Practical past paper questions

1 A student investigated the reaction of aqueous sodium hydroxide with two different acids, acid **C** and acid **D**.
Two experiments were carried out.
Experiment 1
By using a measuring cylinder, 20 cm³ of aqueous sodium hydroxide was poured into a conical flask and the initial temperature of the solution was measured. A burette was filled with acid **C** up to the 0.0 cm³ mark.
5 cm³ of acid **C** was added to the sodium hydroxide in the flask. The temperature of the mixture was measured.
Further 5 cm³ portions of acid **C** were added to the mixture in the flask, stirring with the thermometer until a total volume of 30 cm³ of acid **C** had been added. The temperatures after each 5 cm³ portion had been added were measured.

a Use the thermometer diagrams to record the temperatures in the table of results.

Table of results   [2]

volume of acid **C** added/cm³	thermometer diagrams	temperature/°C
0		
5		
10		
15		
20		

volume of acid **C** added/cm³	thermometer diagrams	temperature/°C
25		
30		

Experiment 2
The burette was emptied and rinsed with water. Experiment 1 was repeated using acid **D**.

b Use the thermometer diagrams to record the temperatures in the table of results.

Table of results   [2]

volume of acid **D** added/cm³	thermometer diagrams	temperature/°C
0		
5		
10		
15		
20		
25		

volume of acid D added/cm³	thermometer diagrams	temperature/°C
30	30 25 20	

c Plot the results for Experiments 1 and 2 on the grid and draw two smooth line graphs. Clearly label your graphs. [6]

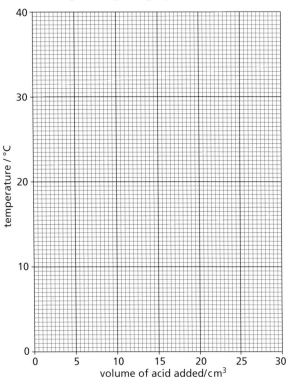

d From your graph, deduce the temperature of the mixture when 3 cm³ of acid C reacted with sodium hydroxide in Experiment 1.
Show clearly **on the graph** how you worked out your answer. [2]

e i Which experiment produced the larger temperature change? [1]
  ii Suggest why the temperature change is greater in this experiment. [2]

f Why was the burette rinsed with water in Experiment 2? [1]

g Predict the temperature of the reaction mixture in Experiment 2 after 1 hour. Explain your answer. [2]

[Total: 18]

(Cambridge IGCSE Chemistry 0620 Paper 61 Q4 June 2010)

2 Solid **E** was analysed. **E** was an aluminium salt. The tests on the solid and some of the observations are in the following table. Complete the observations in the table.

tests	observations
tests on solid **E**	
a Appearance of solid **E**	white crystalline solid
b A little of solid **E** was heated in a test-tube.	colourless drops of liquid formed at the top of the tube
c A little of solid **E** was dissolved in distilled water. The solution was divided into four test-tubes and the following tests were carried out.	
(i) To the first test-tube of solution, drops of aqueous sodium hydroxide were added. Excess sodium hydroxide was then added to the test-tube.	. . . . [3]
(ii) Test (i) was repeated using aqueous ammonia solution instead of aqueous sodium hydroxide.	. . . . [2]
(iii) To the third test-tube of solution, dilute hydrochloric acid was added, followed by barium chloride solution.	no reaction
(iv) To the fourth test-tube of solution, aqueous sodium hydroxide and aluminium powder were added. The mixture was heated.	effervescent pungent gas given off turned damp litmus paper blue

d What does test **(b)** tell you about solid **E**? [1]
e Identify the gas given off in test (c)(iv). [1]
f What conclusions can you draw about solid **E**? [2]

[Total: 9]

(Cambridge IGCSE Chemistry 0620 Paper 61 Q5 June 2010)

3 Hydrated cobalt chloride crystals, $CoCl_2.6H_2O$, were heated in the apparatus shown below.

a Indicate on the diagram, using an arrow, where heat is applied. [1]

b The crystals change colour from _____ to _____ . [1]

c What is the purpose of the ice? [1]

d Why is the tube open at point **W**? [1]

[Total: 4]

(Cambridge IGCSE Chemistry 0620 Paper 61 Q6 June 2010)

4 Ethene can be prepared by passing ethanol vapour over hot aluminium oxide.

a Complete the boxes to show the chemicals used. [1]

b Show on the diagram with two arrows where the heat is applied. [2]

c Why must the delivery tube be removed from the water before the heating is stopped? [2]

[Total: 5]

(Cambridge IGCSE Chemistry 0620 Paper 61 Q1 November 2010)

5 The following instructions were used to prepare magnesium sulfate crystals, $MgSO_4.7H_2O$.

Step 1 Measure $50\,cm^3$ of dilute sulfuric acid into a beaker and warm the solution.

Step 2 Using a spatula, add some magnesium oxide and stir the mixture. Continue adding the magnesium oxide until excess is present.

Step 3 Separate the excess magnesium oxide from the solution of magnesium sulfate.

Step 4 Heat the solution until crystals form. Obtain the crystals and dry them.

a Why is the sulfuric acid warmed? [1]

b How would you know when excess magnesium oxide is present in **Step 2**? [1]

c What method is used in **Step 3**? [1]

d Why must care be taken when drying the crystals in **Step 4**? [1]

e Explain how the method would differ if magnesium carbonate was used instead of magnesium oxide. [2]

[Total: 6]

(Cambridge IGCSE Chemistry 0620 Paper 61 Q2 November 2010)

6 A concentrated solution of sodium chloride was electrolysed using the apparatus below.

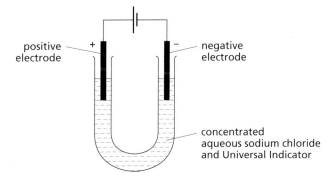

One observation noted was that the Universal Indicator turned purple at the negative electrode.

a What observation would be made at **both** electrodes? [1]

b Why did the indicator turn purple at the negative electrode? [1]

c **i** Name the product formed at the positive electrode. [1]

ii Suggest the effect of this product on the Universal Indicator. [1]

[Total: 4]

(Cambridge IGCSE Chemistry 0620 Paper 61 Q6 November 2010)

7 Two solids, **S** and **V**, were analysed. **S** was copper(II) oxide.
The tests on the solids, and some of the observations, are in the following table.
Complete the observations in the table. Do not write any conclusions in the table.

test	observation
tests on solid **S**	
a Appearance of solid **S**	black solid
b Hydrogen peroxide was added to solid **S** in a test-tube.	slow effervescence
A glowing splint was inserted into the tube.	splint relit
c Dilute sulfuric acid was added to solid **S** in a test-tube. The mixture was heated to boiling point.	blue solution formed
The solution was divided into three equal portions into test-tubes.	
(i) To the first portion of the solution, excess sodium hydroxide was added.	. . . . [1]
(ii) To the second portion of the solution, about 1 cm of aqueous ammonia solution was added. Excess ammonia solution was then added.	. . . . [2] . . . . [2]
(iii) To the third portion of the solution, dilute hydrochloric acid was added followed by barium chloride solution.	. . . . [2]
tests on solid **V**	
d Appearance of solid **V**	black solid
e Hydrogen peroxide was added to solid **V** in a test-tube.	rapid effervescence
A glowing splint was inserted into the tube.	splint relit

f **i** Compare the reactivity of solid **S** and solid **V** with hydrogen peroxide. [1]

ii Identify the gas given off in test **(e)**. [1]

g What conclusions can you draw about solid **V**? [2]

[Total: 11]

(Cambridge IGCSE Chemistry 0620 Paper 06 Q5 June 2009)

8 Acid–base indicators

Indicators are used to identify acids and bases. Indicators can be obtained from berries and other fruits.

a Plan an experiment to obtain an aqueous solution of an indicator from some berries. [3]

b Plan an experiment to use the indicator solution to show that it is an effective indicator. [3]

[Total: 6]

(Cambridge IGCSE Chemistry 0620 Paper 06 Q6 June 2009)

9 Three unlabelled bottles of chemicals each contained one of the following liquids:

- sodium nitrate dissolved in water;
- pure water;
- hexene.

a Give a test by which you could identify sodium nitrate solution.

test _____

result _____ [2]

b Give a test by which you could identify pure water.

test _____

result _____ [2]

c Give a test by which you could identify hexene.

test _____

result _____ [2]

[Total: 6]

(Cambridge IGCSE Chemistry 0620 Paper 06 Q3 November 2009)

10 Dilute hydrochloric acid was added to excess calcium carbonate in a beaker as shown.

The beaker was placed on a balance and the mass of the beaker and contents recorded every minute.
The results are shown in the table.

mass of beaker and contents/g	184.0	178.0	175.6	174.6	174.0	174.0
time/min	0	1	2	3	4	5

a Plot the results on the grid and draw a smooth line graph. [5]

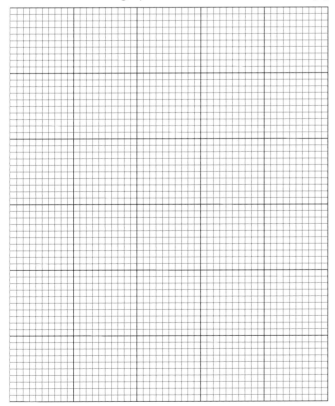

b Use your graph to determine the mass of the beaker and contents after 30 seconds. Show clearly on your graph how you worked out your answer. [2]

c Why does the mass of the beaker and contents decrease? [1]

d Suggest the purpose of the cotton wool. [1]

e After how long did the reaction finish? [1]

f A second experiment was carried out using hydrochloric acid at a lower temperature. On the grid sketch a curve to show the expected results for this experiment. Label this curve **C**. [2]

[Total: 12]

(Cambridge IGCSE Chemistry 0620 Paper 06 Q6 November 2009)

11 A student investigated the reaction between potassium manganate(VII) and a metallic salt solution.
Two experiments were carried out.
Experiment 1

a About 1 cm³ of aqueous sodium hydroxide was added to a little of the salt solution **A** and the observation noted.

observation *green precipitate formed*

b A burette was filled with potassium manganate(VII) solution up to the 0.0 cm³ mark. By using a measuring cylinder, 25 cm³ of solution **A** of the salt was placed into a conical flask. The flask was shaken to mix the contents. The potassium manganate(VII) solution was added to the flask, and shaken to mix thoroughly. Addition of potassium manganate(VII) solution was continued until there was a pale pink colour in the contents of the flask.
Use the burette diagram to record the volume in the table and complete the column.

final reading

Experiment 2

c Experiment **1b** was repeated using a different solution **B** of the salt, instead of solution **A**. Use the burette diagrams to record the volumes in the table and complete the table.

initial reading   final reading

d About 1 cm³ of aqueous sodium hydroxide was added to a little of the solution in the flask and the observation noted.

observation *red-brown precipitate*

Table of results
Burette readings/cm³   [4]

	experiment 1	experiment 2
final reading		
initial reading		
difference		

e i In which experiment was the greatest volume of potassium manganate(VII) solution used?   [1]
  ii Compare the volumes of potassium manganate(VII) solution used in Experiments **1** and **2**.   [2]
  iii Suggest an explanation for the difference in the volumes.   [2]

f Predict the volume of potassium manganate(VII) solution which would be needed to react completely with 50 cm³ of solution **B**.   [2]

g Explain one change that could be made to the experimental method to obtain more accurate results.
  change _____
  explanation _____   [2]

h What conclusion can you draw about the salt solution from
  i experiment **1a**,   [1]
  ii experiment **2d**?   [1]
                    [Total: 15]

(Cambridge IGCSE Chemistry 0620 Paper 06 Q4 June 2008)

12 The colours present in some blackcurrant sweets can be separated by chromatography. The colours are water-soluble dyes. The diagrams show how the colours can be extracted from the sweets.

a Complete the empty boxes to name the pieces of apparatus.   [3]
The apparatus below was used to carry out the chromatography.

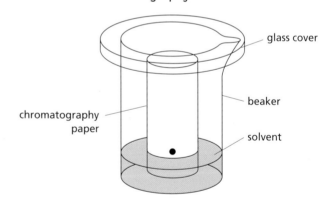

b i Name the solvent used.   [1]
  ii Label, with an arrow, the origin on the diagram.   [1]
c Sketch the chromatogram you would expect if two different colours were present in the sweets.   [1]
                    [Total: 6]

(Cambridge IGCSE Chemistry 0620 Paper 06 Q1 November 2008)

13 Seawater contains sodium chloride and other salts. Plan an experiment to find the mass of salts in 1 dm³ of seawater.
You will be provided with a small bottle of seawater.
You should include details of the method and any apparatus used.
(1 dm³ = 1000 cm³) [6]
[Total: 6]

(Cambridge IGCSE Chemistry 0620 Paper 61
Q6 October/November 2011)

14 Three different liquids **P**, **Q** and **R** were analysed.
P was an aqueous solution of sulfuric acid.
The tests on the liquids and some of the observations are in the following table.
Complete the observations in the table.

tests	observations
a (i) Appearance of the liquids	**P** . . . . [2] **Q** colourless, smell of vinegar **R** colourless, no smell
(ii) The pH of the liquids was tested using Universal indicator paper.	**P** . . . . [2] **Q** pH 5 **R** pH 7
b A piece of magnesium ribbon was added to a little of each liquid. The gas given off by liquid **P** was tested.	**P** . . . . [2] **Q** slow effervescence **R** no reaction
c To a little of liquid **P**, hydrochloric acid and aqueous barium chloride were added.	. . . . [2]
d Liquid **R** was heated to boiling in a test-tube. A thermometer was used to record the constant temperature of the vapour produced.	temperature = 100 °C

e What conclusions can you draw about liquid **Q**? [2]
f Identify liquid **R**. [1]
[Total: 9]

(Cambridge IGCSE Chemistry 0620 Paper 61
Q5 October/November 2011)

Glossary

Acid A substance which dissolves in water, producing $H^+(aq)$ ions as the only positive ions. An acid is a hydrogen ion (H^+) donor.

Activation energy The activation energy, E_a, is the minimum energy that colliding particles must have in order to react.

Addition polymer A polymer formed by an addition reaction. For example, poly(ethene) is formed from ethene.

Addition reaction A reaction in which an atom or group of atoms is added across a carbon–carbon double bond.

Alcohols Organic compounds containing the –OH group. They have the general formula $C_nH_{2n+1}OH$. Ethanol is by far the most important of the alcohols and is often just called 'alcohol'.

Alkali A soluble base which produces $OH^-(aq)$ ions in water.

Alkali metals The six metallic elements in Group I of the Periodic Table.

Alkaline earth metals The six metallic elements in Group II of the Periodic Table.

Alkanes A family of saturated hydrocarbons with the general formula C_nH_{2n+2}. The term 'saturated', in this context, is used to describe molecules that have only single bonds. The alkanes can only undergo substitution reactions in which there is replacement of one atom in the molecule by another atom.

Alkenes A family of unsaturated hydrocarbons with the general formula C_nH_{2n}. The term 'unsaturated', in this context, is used to describe molecules which contain one or more carbon–carbon double bonds. Unsaturated compounds undergo addition reactions across the carbon–carbon double bonds and so produce saturated compounds. The addition of hydrogen across the carbon–carbon double bonds is used to reduce the amount of unsaturation during the production of margarines.

Alloy Generally, a mixture of two or more metals (for example, brass is an alloy of zinc and copper) or of a metal and a non-metal (for example, steel is an alloy of iron and carbon, sometimes with other metals included). They are formed by mixing the molten substances thoroughly. Generally, it is found that alloying produces a metallic substance which has more useful properties than the original pure metals it was made from.

Amino acids These naturally occurring organic compounds possess both an –NH_2 group and a –COOH group on adjacent carbon atoms. There are 20 naturally occurring amino acids, of which glycine is the simplest.

Amphoteric oxide An oxide which can behave as an acid (it reacts with an alkali) or a base (it reacts with an acid), for example, zinc oxide, ZnO, and aluminium oxide, Al_2O_3.

Anaerobic decay Decay that takes place in the absence of oxygen.

Anaerobic respiration Respiration that takes place in the absence of air.

Anhydrous salts Salts which have had their water of crystallisation removed by heating them.

Anions Negative ions; these are attracted to the anode.

Anode The positive electrode. It is positively charged because electrons are drawn away from it.

Artificial fertiliser A substance added to soil to increase the amount of elements such as nitrogen, potassium and phosphorus. This enables crops grown in the soil to grow more healthily and to produce higher yields.

Atmosphere (air) The mixture of gases that surrounds the Earth.

Atmospheric pressure The pressure exerted by the atmosphere on a surface due to the weight of the air above it.

Atom The smallest part of an element that can exist as a stable entity. It has a central nucleus containing neutrons and protons surrounded by electrons in shells. 1 mole of atoms has a mass equal to the relative atomic mass (A_r) in grams.

Atomic mass unit Exactly $\frac{1}{12}$ of the mass of one atom of the most abundant isotope of carbon, carbon-12.

Avogadro's Law Equal volumes of all gases measured under the same conditions of temperature and pressure contain equal numbers of molecules.

Base A substance which neutralises an acid, producing a salt and water as the only products. A base is a hydrogen ion (H^+) acceptor.

Batch process A batch process is one in which each batch goes through one stage of the production process before moving on to the next stage.

Binary compound A substance that contains two elements chemically combined.

Biodegradable This relates to substances which can be broken down in the environment by, for example, bacteria.

Blast furnace A furnace for smelting iron ores, such as hematite (Fe_2O_3) and magnetite (Fe_3O_4), to produce pig (or cast) iron.

Boiling point The temperature at which the pressure of the gas created above a liquid equals atmospheric pressure.

Bond energy An amount of energy associated with a particular bond in a molecular element or compound.

Calculating moles of compounds

$$\text{number of moles} = \frac{\text{mass of compound}}{\text{molar mass of compound}}$$

Calculating moles of elements

$$\text{number of moles} = \frac{\text{mass of element}}{\text{molar mass of element}}$$

Calculating moles of gases 1 mole of any gas occupies 24 dm^3 (litres) at room temperature and pressure (r.t.p.).

$$\frac{\text{number of}}{\text{moles of gas}} = \frac{\text{volume of the gas} \left(\text{in } dm^3 \text{ at r.t.p.}\right)}{24 \ dm^3}$$

Calculating moles of solutions

$$\frac{\text{concentration of a}}{\text{solution (in mol/}dm^3)} = \frac{\text{number of moles of solute}}{\text{volume} \left(\text{in } dm^3\right)}$$

Carboxylic acids A family of organic compounds containing the functional group –COOH. They have the general formula $C_nH_{2n+1}COOH$. The most important and well known of these acids is ethanoic acid, which is the main constituent in vinegar. Ethanoic acid is produced by the oxidation of ethanol.

Catalyst A substance which alters the rate of a chemical reaction and is unchanged at the end of the reaction. It increases the rate of a chemical reaction by providing an alternative reaction path which has a lower activation energy, E_a.

Catalytic converter A device for converting dangerous exhaust gases from cars into less harmful emissions. For example, carbon monoxide gas is converted to carbon dioxide gas.

Catalytic cracking The decomposition of higher alkanes into alkenes and alkanes of lower relative molecular mass. The process involves passing the larger alkane molecules over a catalyst of aluminium and chromium oxides, heated to 550°C.

Cathode The negative electrode. It is negatively charged because an excess of electrons move towards it.

Cations Positive ions; these are attracted to the cathode.

CFC Abbreviation for chlorofluorocarbon, a type of organic compound in which some or all of the hydrogen atoms of an alkane have been replaced by fluorine and chlorine atoms. These substances are generally unreactive but they can diffuse into the stratosphere where they break down under the influence of ultraviolet light. The products of this photochemical process then react with ozone (in the ozone layer). Because of this, their use has been discouraged. They are now being replaced by hydrochlorofluorocarbons (HCFCs).

Chemical change A permanent change in which a new substance is formed.

Chemical formula A shorthand method of representing chemical elements and compounds.

Chromatography A technique employed for the separation of mixtures of dissolved substances.

Combustion A chemical reaction in which a substance reacts rapidly with oxygen with the production of heat and light.

Competition reactions Reactions in which metals compete for oxygen or anions.

Compound A substance formed by the combination of two or more elements in fixed proportions.

Condensation The change of a gas into a liquid. This process is accompanied by the evolution of heat.

Condensation polymer A polymer formed by a condensation reaction (one in which water is produced during polymerisation). For example, nylon is produced by the condensation reaction between 1,6-diaminohexane and hexanedioic acid.

Corrosion The process that takes place when metals and alloys are chemically attacked by oxygen, water or any other substances found in their immediate environment.

Covalent bond A chemical bond formed by the sharing of one or more pairs of electrons between two atoms.

Crystal A solid whose particles are arranged in a definite way giving a specific shape.

Crystallisation The process of forming crystals from a liquid.

Decanting The process of removing a liquid from a solid which has settled or from an immiscible heavier liquid by pouring.

Delocalised Refers to spreading out of electrons within the metal structure. The electrons are not attached to any one particular ion.

Diatomic molecule A molecule containing two atoms, for example, hydrogen, H_2, and oxygen, O_2.

Diffusion The process by which different substances mix as a result of the random motions of their particles.

Displacement reaction A reaction in which a more reactive element displaces a less reactive element from solution.

Distillate The condensed vapour produced from a mixture of liquids on distillation.

Distillation The process of boiling a liquid and then condensing the vapour produced back into a liquid. It is used to purify liquids and to separate mixtures of liquids.

Electrical conductivity A substance shows electrical conductivity if it allows the passage of an electric current through.

Electrode A point where the electric current enters and leaves the electrolytic cell. An inert electrode is usually made of platinum or carbon and does not react with the electrolyte or the substances produced at the electrodes themselves.

Electrolysis The decomposition of an ionic compound, when molten or in aqueous solution, by the passage of an electric current.

Electrolyte A substance which will carry electric current only when it is molten or dissolved.

Electron A sub-atomic particle that has a negative charge equal in magnitude to that of a proton. Electrons are found in electron shells which surround the nucleus of the atom.

Electronic configuration (structure) A shorthand method of describing the arrangement of electrons within the electron shells of an atom.

Electron shells (energy levels) Electrons group together in electron shells which surround the nucleus of the atom. The electron shells can hold differing numbers of electrons.

Electroplating The process of depositing metals from solution in the form of a layer on other surfaces such as metal.

Electrostatic force of attraction A strong force of attraction between opposite charges.

Element A substance which cannot be further divided into simpler substances by chemical methods.

Empirical formula A formula showing the simplest whole number ratio of atoms or ions present in a compound.

Endothermic reaction An endothermic reaction absorbs thermal energy from the surroundings, leading to a decrease in the temperature of the surroundings.

Enthalpy Energy stored in chemical bonds, given the symbol *H*.

Enthalpy change of reaction The transfer of thermal energy during a reaction is called the enthalpy change, ΔH, of the reaction. ΔH is negative for exothermic reactions and positive for endothermic reactions.

Enthalpy of combustion The enthalpy change which takes place when 1 mole of a substance is completely burned in oxygen.

Enthalpy of neutralisation The enthalpy change which takes place when 1 mole of hydrogen ions is completely neutralised.

Enzymes Protein molecules produced in living cells. They act as biological catalysts and are specific to certain reactions. They operate only within narrow temperature and pH ranges.

Esters A family of organic compounds formed by the reaction of an alcohol with a carboxylic acid in the presence of concentrated H_2SO_4.

Eutrophication A process that occurs when fertiliser drains into lakes and rivers, causing algae to multiply rapidly and the water to turn green. It results in fish and other organisms suffocating and dying through lack of oxygen in the water.

Evaporation A process occurring at the surface of a liquid involving the change of state of a liquid into a gas at a temperature below the boiling point. When a solution is heated, the solvent evaporates and leaves the solute behind.

Exothermic reaction An exothermic reaction transfers thermal energy to its surroundings, leading to an increase in the temperature of the surroundings.

Fermentation A series of biochemical reactions brought about by the enzymes in yeast or, more generally, by micro-organisms.

Filtrate The liquid or solution which passes through a filter during filtration.

Filtration The process of separating a solid from a liquid using a fine filter paper which does not allow the solid to pass through.

Flue gas desulfurisation (FGD) The process by which sulfur dioxide gas is removed from the waste gases of power stations by passing them through calcium hydroxide slurry.

Fossil fuels Fuels, such as coal, oil and natural gas, formed from the remains of plants and animals.

Fractional distillation A distillation technique used to separate a mixture of liquids that have different boiling points.

Fuel cells In a hydrogen-oxygen fuel cell, hydrogen and oxygen are used to produce a voltage, and water is the only product.

Functional group The atom or group of atoms responsible for the characteristic reactions of a compound. For example, the alcohol functional group is −OH.

Giant covalent structure A substance containing thousands of atoms per molecule.

Giant ionic lattice A lattice of positive and negative ions held together by the electrostatic forces of attraction between ions.

Group A vertical column of the Periodic Table containing elements with the same number of electrons in their outer electron shell. Elements in a particular group show trends in their properties, both chemical and physical.

Haber process The chemical process by which ammonia, NH_3, is made in very large quantities from nitrogen and hydrogen.

Hall–Héroult cell The electrolysis cell in which aluminium is extracted from purified bauxite dissolved in molten cryolite at 900°C. This cell has both a graphite anode and a graphite cathode.

Halogens The elements found in Group VII of the Periodic Table.

Homologous series A series of compounds which possess the same functional group and have the same general formula. The members of the series differ from one member to the next by a $-CH_2-$ unit and show a trend in physical properties as well as having similar chemical properties.

Hydrated salts Salts which contain water molecules in their crystal structures.

Hydrocarbon A compound which contains atoms of carbon and hydrogen only.

Hydrogen bond A weak intermolecular force (bond) that occurs between water molecules because the bonds within the molecules are polar. It is the strongest of these weak forces.

Immiscible When two liquids form two layers when mixed together, they are said to be immiscible.

Indicator A substance used to show whether a substance is acidic or alkaline (basic), for example, thymolphthalein.

Inert electrode These are electrodes that do not react with the electrolyte or the products of electrolysis. Examples of inert electrodes are carbon and platinum.

Insoluble If a solute does not dissolve in a solvent, it is said to be insoluble.

Intermolecular bonds These are attractive forces which form between neighbouring molecules.

Intramolecular bonds Forces which act within a molecule, for example, covalent bonds.

Ion An atom or group of atoms which has either lost one or more electrons, making it positively charged, or gained one or more electrons, making it negatively charged.

Ionic bond A strong electrostatic force of attraction between oppositely charged ions.

Ionic equation The simplified equation of a reaction which we can write if the chemicals involved are ionic substances.

Ionisation The process whereby an atom gains or loses an electron(s) to become an ion.

Isomers Compounds which have the same molecular formula but different structural arrangements of the atoms.

Isotopes These are different atoms of the same element that have the same number of protons but different numbers of neutrons. Isotopes of the same element have the same chemical properties because they have the same number of electrons and therefore the same electronic configuration.

Kinetic particle theory A theory which accounts for the properties of matter in terms of its constituent particles.

Lattice A regular three-dimensional arrangement of atoms/ ions in a crystalline solid.

Law of constant composition Compounds always have the same elements joined together in the same proportion.

Locating agent A substance used to locate, on a chromatogram, the separated parts of a mixture in chromatography.

Mass number (nucleon number) Symbol A. The total number of protons and neutrons found in the nucleus of an atom.

Matter Anything which occupies space and has a mass.

Melting point The temperature at which a solid begins to turn into a liquid. Pure substances have a sharp melting point.

Metal extraction The method used to extract a metal from its ore; it depends on the position of the metal in the reactivity series.

Metallic bond An electrostatic force of attraction between the 'sea' of electrons and the regular array of positive metal ions within a solid metal.

Metalloid (semi-metal) An element with properties between those of metals and non-metals, for example boron and silicon.

Metals A class of chemical elements which have a characteristic lustrous appearance and are good conductors of heat and electricity.

Miscible When two liquids form a homogeneous layer when mixed together, they are said to be miscible.

Mixture Two or more substances mixed together that can be separated by physical means.

Molar mass The mass of 1 mole of a compound. It has units of g/mol.

Mole The amount of substance which contains 6.02×10^{23} atoms, ions or molecules. This number is called Avogadro's constant.

Molecular formula A formula showing the actual number and type of different atoms of each element present in one molecule of a compound.

Molecule A group of atoms covalently bonded together.

Monatomic molecule A molecule which consists of only one atom, for example neon or argon.

Monomer A simple molecule, such as ethene, which can be polymerised.

Neutralisation The process in which the acidity or alkalinity of a substance is destroyed. Destroying acidity means removing $H^+(aq)$ by reaction with a base, carbonate or metal. Destroying alkalinity means removing the $OH^-(aq)$ by reaction with an acid: $H^+(aq) + OH^-(aq) \rightarrow H_2O(l)$.

Neutron An uncharged sub-atomic particle present in the nuclei of atoms.

Noble gases The unreactive gases found in Group 0 of the Periodic Table.

Non-metals A class of chemical elements that are typically poor conductors of heat and electricity.

Non-renewable energy sources Sources of energy, such as fossil fuels, which take millions of years to form and which we are using up at a rapid rate.

Nucleus Found at the centre of the atom, it consists of protons and neutrons.

Oil refining The general process of converting the mixture that is collected as petroleum into separate fractions. The fractions are separated from the petroleum mixture by fractional distillation.

Ore A naturally occurring mineral from which a metal can be extracted.

Organic chemistry The branch of chemistry concerned with compounds of carbon found in living organisms.

Organic compounds Substances whose molecules contain one or more carbon atoms covalently bonded with another element (including hydrogen, nitrogen, oxygen, the halogens as well as phosphorus, silicon and sulfur).

Oxidation Gain of oxygen.

Oxidation The process of oxygen gain or loss of electrons. Also an increase in oxidation number.

Oxidising agent A substance that oxidises another substance and is itself reduced.

Oxidising agent A substance that oxidises another substance and is itself reduced during a redox reaction. It will cause an increase in oxidation number.

Particulates Very small particles found in the atmosphere, such as certain types of smoke emitted from diesel engines, as well as dust.

Percentage yield

$$\text{percentage yield} = \frac{\text{actual yield}}{\text{theoretical yield}} \times 100$$

Periodic Table A table of elements arranged in order of increasing proton number to show the similarities of the chemical elements with related electronic configuration.

Periods Horizontal rows of the Periodic Table. Within a period, the atoms of all the elements have the same number of occupied electron shells but have an increasing number of electrons in the outer electron shell.

Petroleum Petroleum is a mixture of hydrocarbons. It is often referred to as crude oil.

pH scale A scale running from 0 to 14, used for expressing the acidity or alkalinity of a solution.

Photosynthesis The chemical process by which green plants synthesise their carbon compounds from atmospheric carbon dioxide using light as the energy source and chlorophyll as the catalyst.

Pollution The modification of the environment caused by human influence. It often renders the environment harmful and unpleasant to life. Atmospheric pollution is caused by gases such as sulfur dioxide, carbon monoxide and nitrogen oxides being released into the atmosphere by a variety of industries and also by the burning of fossil fuels. Water pollution is caused by many substances, such as those found in fertilisers and in industrial effluent.

Polymer A substance possessing very large molecules consisting of repeated units or monomers. Polymers therefore have a very large relative molecular mass.

Polymerisation The chemical reaction in which molecules (monomers) join together to form a polymer.

Polyamide A condensation polymer, such as nylon, that contains the amide link, –NHOC–.

Proteins Natural polyamides formed from amino acids by condensation reactions.

Proton A sub-atomic particle that has a positive charge equal in magnitude to that of an electron. Protons occur in all nuclei.

Proton number (atomic number) The number of protons in the nucleus of an atom. Symbol Z. The proton or atomic number is also equal to the number of electrons present in an atom.

Pure substances These are substances that are made of only one type of atom, compound or one type of molecule (a group of atoms bonded together).

Purity The measure of whether a substance is pure.

Radioactive A property of unstable isotopes. They disintegrate spontaneously to give off energy and one or more types of radiation.

Radioisotope An isotope that gives off energy called radiation.

Reaction rate A measure of the change which happens during a reaction in a single unit of time. It may be affected by the following factors: surface area of the reactants, concentration of the reactants, the pressure of gases, the temperature at which the reaction is carried out and the use of a catalyst.

Reactivity series of metals An order of reactivity, giving the most reactive metal first, based on results from experiments with oxygen, water and dilute hydrochloric acid.

Redox reaction A reaction which involves simultaneous reduction and oxidation.

Reducing agent A substance which reduces another substance and is itself oxidised.

Reducing agent A substance which reduces another substance and is itself oxidised during a redox reaction. Also it will cause a decrease in oxidation number.

Reduction The loss of oxygen.

Reduction The process of oxygen loss, electron gain. Also a decrease in oxidation number.

Relative atomic mass Symbol A_r.

$$A_r = \frac{\text{average mass of isotope of the element}}{\frac{1}{12} \times \text{mass of 1 atom of carbon-12}}$$

Relative formula mass (RFM) The sum of the relative atomic masses of all those elements shown in the formula of the substance. This is often referred to as the relative molecular mass (M_r).

Renewable energy Sources of energy which cannot be used up or which can be made at a rate faster than the rate of use.

Residue The substance that remains after evaporation, distillation or filtration have taken place.

Reversible reaction A chemical reaction which can go both forwards and backwards. This means that once some of the products have been formed, they will undergo a chemical change once more to re-form the reactants. The reaction from left to right in the equation is known as the forward reaction and the reaction from right to left is known as the reverse reaction.

R_f value The ratio of the distance travelled by the solute to the distance travelled by the solvent in chromatography.

$$R_f = \frac{\text{distance travelled by solute}}{\text{distance travelled by solvent}}$$

Rust A loose, orange-brown, flaky layer of hydrated iron(III) oxide found on the surface of iron or steel. The conditions necessary for rusting to take place are the presence of oxygen and water. The rusting process is encouraged by other substances such as salt. It is an oxidation process.

Rust prevention To prevent iron rusting, it is necessary to stop oxygen and water coming into contact with it. The methods employed include painting, oiling/greasing, coating with plastic, plating, galvanising and sacrificial protection.

Saturated hydrocarbon A type of hydrocarbon molecule in which the molecule has the maximum possible number of hydrogen atoms and so has no double bonds.

Saturated solution A solution containing the maximum concentration of a solute dissolved in the solvent at a given temperature.

Simple molecular substance A substance that possesses between one and a few hundred atoms per molecule.

Smog Caused by atmospheric pollutants that can react in sunlight with other substances to produce a hazy, harmful atmosphere.

Solids, liquids and gases The three states of matter to which all substances belong.

Soluble If a solute dissolves in a solvent, it is said to be soluble.

Solute The substance that is dissolved in a solvent.

Solution A liquid mixture composed of two or more substances.

Solubility The solubility of a solute in a solvent at a given temperature is the number of grams of the solute which can dissolve in 100 g of solvent to produce a saturated solution at that temperature.

Solvent A substance that dissolves a solute to form a solution.

$$\text{solute} + \text{solvent} \xrightarrow{\text{dissolves}} \text{solution}$$

Strong acid An acid which produces a high concentration of $H^+(aq)$ ions in water solution, for example hydrochloric acid.

Substitution reaction A reaction in which an atom or group of atoms is replaced by another atom or group of atoms.

Supercooled liquid One which has cooled below its freezing point without solidification.

Thermal cracking The decomposition of higher alkanes to alkenes of lower relative molecular mass at high temperatures, 800–850°C.

Titrant A solution of known concentration that is added (titrated) to another solution to determine the concentration of a second chemical substance.

Titration A method of volumetric analysis in which a volume of one reagent (for example, an acid) is added to a known volume of another reagent (for example, an alkali) slowly from a burette until an end-point is reached. If an acid and alkali are used, then an indicator is used to show that the end-point has been reached.

Toxic Harmful to health, even deadly.

Transition elements The elements found in the centre of the Periodic Table, between Groups II and III.

Trend A term used by chemists to describe how a particular property changes across a period or down a group of the Periodic Table.

Unsaturated hydrocarbon A hydrocarbon molecule that possess at least one carbon–carbon double or triple covalent bond.

Water of crystallisation Water incorporated into the structure of substances as they crystallise, for example, in copper(II) sulfate pentahydrate ($CuSO_4.5H_2O$).

Weak acid An acid which produces a low concentration of $H^+(aq)$ in water solution, for example, ethanoic acid. It is only partially ionised.

X-ray diffraction A technique used to study crystal structures.

Index

Note: page numbers in *italics* indicate references to suggestions for practical work.

O

oil refining 88–9
OIL RIG mnemonic 73
oils 213
ores 157–8
organic compounds 88, 188
see also alcohols; alkanes; alkenes;
   carboxylic acids; esters;
   plastics; polymers
oxidation 15, 31, 71
oxidation numbers 35–7
   transition elements 147
oxides 14, 121, 160
oxidising agents 15, 36–7, 208
oxygen
   in atmosphere 178
   electronic configuration 26
   physical properties 4, 11
   reaction with metals 151, 152
   size of atoms 12
   testing for 236
   uses of 184
oxygen molecules 42
ozone hole 192–3

P

paper chromatography 230–2
Paris Accord 183, 193
particulates 183
   atmospheric pollution 179, 181
pentane 189
peptide links 215, 216
percentage composition 65
percentage purity 65
percentage yield 65
Periodic Table 135
   development of 134–6
   and electronic configuration 137
   Group I (alkali metals) 137–41
   Group II (alkaline earth metals)
      140–1
   groups and periods 136
   Group VII (halogens) 141–5
   Group VIII (noble gases) 145
   hydrogen 147
   transition elements 145–7
   trends 137, 139
Perspex 200
PET (poly(ethylene terephthalate))
   215–16
petroleum 87
   fractional distillation 88–9
   percentages of fractions 196
pewter 166
pH meters 118
phosphates
   in fertilisers 175
   pollution 173

photochemical smogs 180
photodegradable plastics 201, 202
photosynthesis 178
pH scale 117
physical changes 4, 100
physical properties 11
   alcohols 207
   alkali metals 138
   alkaline earth metals 140
   alkanes 190
   alkenes 194
   covalent compounds 43–4
   graphite and diamond 47
   halogens 141
   ionic compounds 34
   metals 50–1, 150
   non-metals 150
   transition elements 145
pipettes 222, 223
plastics 43, 198–200
   electroplating 83
plastic waste 173–4, 201–2
platinum 164
polar molecules 173, 207
pollution
   acid rain *182*
   atmospheric 179–82
   bauxite 74
   from plastics 201–2
   of water 173–4
polyamides 215
   natural 216–17
polyesters 215
poly(ethene) 198, 199, 200
polymers
   addition 198–200
   condensation 214–17
poly(propene) 200
polystyrene 200
potassium 137–8, 153, 164
   electronic configuration 26, 140
   in fertilisers 175
   flame colour 234
   reaction with water 138, 139
potassium hydroxide 59
   reaction with nitric acid *127–8*
potassium iodide 37
potassium manganate(VII) 37, 208
potassium nitrate, preparation
   of *127–8*
power stations 91
precipitation 124
precision 222
pressure
   effect on Haber process 110–11
   effect on reaction rates 105
   effects on volume of a gas 6
products 17
propane 189, 191

propanoic acid 211
propanol 207
propene 194, 195
propyl methanoate 213
proteins 216–17
proton number (Z) 21–2
protons 20–2
PTFE (polytetrafluoroethene)
   199–200
pure substances 14
purity 232
   assessment of 233
PVC (polyvinyl chloride) 199–200

Q

qualitative analysis 233, *237*
   appearance or smell 233–4
   flame colours 234
   tests for aqueous anions 236
   tests for aqueous cations 234–5
   tests for gases 236
quartz 44

R

radioisotopes 24
radiotherapy 24
radium 140
radon 145
Ramsay, Sir William 145
Rayleigh, Lord (John William Strutt) 145
reactants 17
reaction pathway diagrams 92, 94
reaction rates 100–1
   effect of catalysts 106–7
   effect of concentration 104–5
   effect of gas pressure 105
   effect of surface area 102–4, *103*
   effect of temperature 105–6
   measurement 102
reactivity series 151, 152
recycling
   aluminium 74
   copper 80
   metals 160–1
   plastics 201
redox reactions 15, 31, 36–7
reducing agents 15, 36–7
reduction 15, 31, 71
relative atomic mass (A_r) 25, 54
relative formula mass (M_r) 55, 56
relative molecular mass (M_r) 54–5
renewable sources of energy
   91, 184
residue 225
reversible reactions 108–9, 119, 197
   equilibrium 110
R_f (retardation factor) values 232
rusting 161, 162
rust prevention 161–4